# History of Rocketry and Astronautics
AAS History Series, Volume 12
International Academy of Astronautics
History Symposia

**Front Cover Illustration:**

Dr. Theodore von Kármán (May 11, 1881 - May 6, 1963) - Aeronautics and Astronautics Pioneer.

# History of
# Rocketry and Astronautics

Proceedings of the Seventeenth History Symposium of the
International Academy of Astronautics

Budapest, Hungary, 1983

John L. Sloop, Volume Editor

R. Cargill Hall, Series Editor

# AAS History Series, Volume 12
A Supplement to Advances in the Astronautical Sciences

IAA History Symposia, Volume 7

Copyright 1991

by

AMERICAN ASTRONAUTICAL SOCIETY

AAS Publications Office
P.O. Box 28130
San Diego, California 92198

Affiliated with the American Association for the Advancement of Science
Member of the International Astronautical Federation

*First Printing 1991*

ISSN 0730-3564

ISBN 0-87703-332-3 (Hard Cover)
ISBN 0-87703-333-1 (Soft Cover)

*Published for the American Astronautical Society
by Univelt, Inc., P.O. Box 28130, San Diego, California 92198*

Printed and Bound in the U.S.A.

## FOREWORD

The roots of rocketry and the related discipline of astronautics can be traced backward more than one-thousand years. Yet "the history" of rocketry and astronautics is essentially contemporary: the most important of its scientific and technical applications have transpired within living memory. Many of us have watched this history unfold and observed the profound effects it has had on strategic and tactical warfare, international politics, society, and the economy. The papers that appear in this volume tell us about this history; about the lessons learned and forgotten, about the dramatic expansion in the commercial communication satellite business, and about the simultaneous contraction and demise of a once-dominant American rocket company.

History, among other things, is our collective experience; it is also a hard teacher. Although a knowledge of history does not guarantee our decisions of business or state will be error-free, it can help us avoid repeating mistakes. Those who practice the engineering and science of astronautics and fail to reflect on their history court ignorance, which will one day surely imperil them. Worse still are those who selectively interpret history to suit preset purposes; they even more surely court catastrophe. Of recent lessons, we may at least hope that other nations will not repeat America's misguided attempt in the 1980s to operate all of its civil, commercial, and military space programs exclusively on a fleet of four space shuttles.

The exploration and use of outer space for the benefit of mankind is a compelling enterprise. Explore for yourself the pages that follow.

                                            R. Cargill Hall
                                            Series Editor
                                            Office of Air Force History
                                            Washington, D.C.

# AAS HISTORY SERIES

**Volume 1**     *Two Hundred Years of Flight in America: A Bicentennial Survey*, Edited by Eugene M. Emme, 1977, 326p, Third Printing 1981, Hard Cover $35; Soft Cover $25; special price for classroom text or bulk purchase.

**Volume 2**     *Twenty-Five Years of the American Astronautical Society: Historical Reflections and Projections*, 1954-1979, Edited by Eugene M. Emme, 1980, 248p, Hard Cover $25; Soft Cover $15.

**Volume 3**     *Between Sputnik and the Shuttle: New Perspectives on American Astronautics*, 1957-1980, Edited by Frederick C. Durant, III, 1981, 350p, Hard Cover $40; Soft Cover $30.

**Volume 4**     *The Endless Space Frontier: A History of the House Committee on Science and Astronautics*, By Ken Hechler, Abridged and edited by Albert E. Eastman, 1982, 460p, Hard Cover $45; Soft Cover $35.

**Volume 5**     *Science Fiction and Space Futures: Past and Present*, Edited by Eugene M. Emme, 1982, 278p, Hard Cover $35; Soft Cover $25.

**Volume 6**     *First Steps Toward Space*, Edited by Frederick C. Durant, III and George S. James, 1986, 318p, Hard Cover $45; Soft Cover $35.

**Volume 7**     *History of Rocketry and Astronautics*, Edited by R. Cargill Hall, 1986, Part I, 250p, Part II, 502p, sold as a set, Hard Cover $100; Soft Cover $80.

**Volume 8**     *History of Rocketry and Astronautics*, Edited by Kristan R. Lattu, 1989, 368p, Hard Cover $50; Soft Cover $35.

**Volume 9**     *History of Rocketry and Astronautics*, Edited by Frederick I. Ordway, III, 1989, 330p, Hard Cover $50; Soft Cover $35.

**Volume 10**     *History of Rocketry and Astronautics*, Edited by Å. Ingemar Skoog, 1990, 330p, Hard Cover $60; Soft Cover $40

**Volume 11**     *History of Rocketry and Astronautics*, Edited by Mitchell R. Sharpe, 1991, Hard Cover $60; Soft Cover $40

**Volume 12**     *History of Rocketry and Astronautics*, Edited by John L. Sloop, 1991, 252p, Hard Cover $60; Soft Cover $40

These volumes are available to individual members of space and astronautical societies and students at half the list price.

*Order from Univelt, Incorporated, P.O. Box 28130, San Diego, California 92198*

# PREFACE

This volume of the American Astronautical Society's history series contains fifteen papers presented at the 17th History Symposium of the International Academy of Astronautics. The symposium was held during the 34th Congress of the International Astronautical Federation that convened in Budapest, Hungary, October 10-15, 1983. Regrettably, a paper by G. Yu. Masing of the Soviet Union, "On Some Regularities in the Development of Solid-Propellant Rocket Engines," is missing from these pages. Repeated attempts to obtain the six figures and two tables representing the theoretical and experimental data on which the Masing paper was based proved unsuccessful.

The remaining papers are divided among five parts, following the format of earlier volumes. These papers cover a variety of subjects and activities in seven countries, including war rockets as early as 1377, rocket society activities in three countries, theoretical and practical contributions to rocket propulsion technology in four countries, the early evolution of communication satellites, initial nuclear rocket experiments, the history of a rocket firm, and even a brief, albeit facile description of a contemporary space camp.

Whenever practical for the convenience of the reader, values in English units are followed by values in the International System of Units (SI).

The editor wishes to thank the authors who responded to his requests for illustrations. He is indebted especially to Frederick I. Ordway, III, who served as mentor on this novel assignment, to R. Cargill Hall, the Series Editor, and Horace Jacobs, the publisher.

John L. Sloop
**Volume Editor**
Bethesda, Maryland

# PUBLICATIONS OF THE
# AMERICAN ASTRONAUTICAL SOCIETY

Following are the principal publications of the American Astronautical Society:

JOURNAL OF THE ASTRONAUTICAL SCIENCES (1954 -   )
Published quarterly and distributed by AAS Business Office, 6352 Rolling Mill Place, Suite #102, Springfield, Virginia 22152. Back issues available from Univelt, Inc., P.O. Box 28130, San Diego, California 92198.

SPACE TIMES (1986 -   )
Published bi-monthly and distributed by AAS Business Office, 6352 Rolling Mill Place, Suite #102, Springfield, Virginia 22152.

AAS NEWSLETTER (1962 - 1985)
Incorporated in *Space Times*. Back issues available from AAS Business Office, 6352 Rolling Mill Place, Suite #102, Springfield, Virginia 22152.

ASTRONAUTICAL SCIENCES REVIEW (1959 - 1962)
Incorporated in *Space Times*. Back issues still available from Univelt, Inc., P.O. Box 28130, San Diego, California 92198.

ADVANCES IN THE ASTRONAUTICAL SCIENCES (1957 -   )
Proceedings of major AAS technical meetings. Published and distributed for the American Astronautical Society by Univelt, Inc., P.O. Box 28130, San Diego, California 92198.

SCIENCE AND TECHNOLOGY SERIES (1964 -   )
Supplement to *Advances in the Astronautical Sciences*. Proceedings and monographs, most of them based on AAS technical meetings. Published and distributed for the American Astronautical Society by Univelt, Inc., P.O. Box 28130, San Diego, California 92198.

AAS HISTORY SERIES (1977 -   )
Supplement to *Advances in the Astronautical Sciences*. Selected works in the field of aerospace history under the editorship of R. Cargill Hall. Published and distributed for the American Astronautical Society by Univelt, Inc., P.O. Box 28130, San Diego, California 92198.

AAS MICROFICHE SERIES (1968 -   )
Supplement to *Advances in the Astronautical Sciences*. Consists principally of technical papers not included in the hard-copy volume. Published and distributed for the American Astronautical Society by Univelt, Inc., P.O. Box 28130, San Diego, California 92198.

Subscriptions to the *Journal of the Astronautical Sciences* and the *Space Times* should be ordered from the AAS Business Office. Back issues of the *Journal* and all books and microfiche should be ordered from Univelt, Incorporated.

# CONTENTS

                                                                                              Page

**FOREWORD, R. Cargill Hall, Series Editor**      v

**PREFACE, John L. Sloop, Volume Editor**      vii

**PART I -- EARLY SOLID-PROPELLANT ROCKETRY**      1

Chapter 1: A Study of Early Korean Rockets (1377-1600)
    Chae, Yeon Seok . . . . . . . . . . . . . . 3

**PART II -- ROCKETRY AND ASTRONAUTICS: CONCEPTS, THEORIES, AND ANALYSES**      17

Chapter 2: Leonhard Euler's Importance for Aerospace Sciences - on the Occasion of the Bicentenary of his Death
    Werner Schulz . . . . . . . . . . . . . . 19

**PART III -- THE DEVELOPMENT OF LIQUID- AND SOLID-PROPELLANT ROCKETS, 1880-1945**      29

Chapter 3: The Founding of the Jet Propulsion Research Institute and the Main Fields of its Activity
    B. V. Rauschenbach . . . . . . . . . . . . 31

Chapter 4: The British Interplanetary Society: The First Fifty Years (1933-1983)
    G. V. E. Thompson and L. R. Shepherd . . . . . . . . . 37

Chapter 5: Liquid Propellant Rocket Development by the U.S. Navy during World War II: A Memoir
    Robert C. Traux . . . . . . . . . . . . . . 57

Chapter 6: Some Vignettes from an Early Rocketeer's Diary: A Memoir
    Bernard Smith as told to Frederick I. Ordway, III . . . . . . . 69

|  | Page |
|---|---|
| Chapter 7: Contribution of the Romanian Inventor Alexandru Churcu to the Development of Theoretical and Practical Reactive Motion in the 19th Century | |
| Florin Zăgănescu, Rodica Burlacu and I. M. Stefan | 85 |

## PART IV -- ROCKETRY AND ASTRONAUTICS AFTER 1945     93

Chapter 8: Communication Satellites: The Experimental Years
    Burton I. Edelson . . . . . . . . . . . 95

Chapter 9: Project Rover: The United States Nuclear Rocket Program
    James A. Dewar . . . . . . . . . . . 109

Chapter 10: A Comparative Study of the Evolution of Manned and Unmanned Spaceflight Operations
    Kristan Lattu . . . . . . . . . . 125

Chapter 11: Reaction Motors Division of Thiokol Chemical Corporation: An Operational History, 1958-1972 (Part II)
    Frederick I. Ordway, III . . . . . . . . . . . 137

Chapter 12: Reaction Motors Division of Thiokol Chemical Corporation: A Project History, 1958-1972 (Part III)
    Frank H. Winter . . . . . . . . . . . 175

Chapter 13: Pages from the History of the Hungarian Astronautical Society
    István György Nagy . . . . . . . . . 203

Chapter 14: United States Space Camp at the Alabama Space and Rocket Center
    Edward O. Buckbee and Lee Sentell . . . . . . . . . . . 209

## PART V -- PIONEERS OF ROCKETRY AND ASTRONAUTICS     215

Chapter 15: A Life Devoted to Astronautics: Dr. Olgierd Wołczek (1922-1982)
    M. Subotowicz . . . . . . . . . . . 217

## APPENDIX     233

Biographical Sketches of the Authors . . . . . . . . . . . 235

## INDEX     239

Numerical Index . . . . . . . . . . . 241

Author Index . . . . . . . . . . . 242

# Part I

# EARLY SOLID-PROPELLANT ROCKETRY

# Chapter 1

# A STUDY OF EARLY KOREAN ROCKETS (1377-1600)[*]

## Chae, Yeon Seok[†]

The first Korean rocket was fired between 1377-1389 and began the Korean development of rockets as a tactical weapon. Although Korea has successfully demonstrated the use of rockets as firearms in the 15th Century, there had been no effort to present the historical development of the early Korean rockets in a paper which will be useful to both historians and scientists. The book entitled *Kuk Cho Ore Sorye* (1474) in the Korean language provided extensive rocket system descriptions; however it required considerable research to interpret them. This paper is the first study of early Korean rockets and launchers. The major effort in this study is directed toward the development of design concepts and details of early Korean rockets. Also, to substantiate support of the historical data presented, some versions of the early Korean rockets were made according to their specifications and fired successfully by the author in 1981. Modern working drawings were made from the ancient descriptions, and were used to construct and fire modern copies of the ancient weapons.[1]

The oldest book in the field of firearms in Korea is *Kuk Cho Ore Sorye*[2], published in 1474. The chapter "Firearms Illustrations" contained figures and very detailed descriptions of 23 kinds of firearms that were developed between 1448-1452, excluding a description used in the process of manufacturing of black powder in Korea. "The Introduction and Development of Firearms in Korea (1356-1474)" was written by Ho, Son-Do[3], who has been studying Korean ancient firearms since early 1960. He has written several papers on Korean firearms except the rocket propelled arrows.[4] He mainly used the *Cho Son Wang Cho Silok*[5] for his study, which is an important source of history of Korean firearms. Other papers on Korean firearms are those by Jeon[6] and Boots.[7]

Ancient Korean rockets have otherwise received little attention even though Koreans have used gunpowder since 1377 and have made many kinds of firearms for several hundred years.[8]

---

[*] Presented at the Seventeenth History Symposium of the International Academy of Astronautics, Budapest, Hungary, 1983.

[†] Graduate Student, Aerospace Engineering, Mississippi State University, Mississippi State, Mississippi; Student Member AIAA.

The first black powder in Korea was made by Choi Mu-Son in 1377. During the third year of Shin Yu (1377), the office of a Hwa-tong-do-gam, which means a general bureau of gunpowder artillery, was first established. This was suggested by a certain Choi, Mu-Son, who lived in the same city with Lee Yuan, known for getting saltpeter for the Yuan army. Choi Mu-Son learned from him the procedure for preparing gunpowder. He trained his own workmen, and proposed the above-mentioned establishment of a general bureau of gun powder artillery.

## THE FIRST ROCKET IN KOREA

Choi Mu-Son made many kinds of firearms and gunpowder, according to *Koryo Sa* and *Cho Son Wang Cho Silok*. The running-fire and fire arrow are among the firearms which were made by Choi Mu-Son between 1377-1391. Some types of the Chinese fire-arrows were rocket-propelled arrows, but Choi's fire-arrow was not a rocket-propelled arrow. It was only an incendiary arrow shot from a bow, (Figure 1).

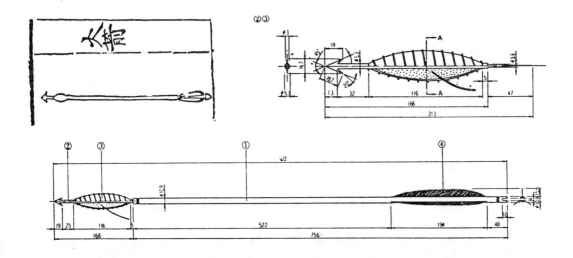

**Figure 1**  Fire-arrow's drawing in the Kuk Cho Ore Sorye (1474) and new plan in the Early Firearms in Korea (1377-1600), 1) arrow shaft, 2) arrow head, 3) black powder, 4) fins.

The structure of the Choi's fire-arrow was as follows, according to the "Firearms Illustration" of *Kuk Cho Ore Sorye*, which used Korean measures for description of firearms. The author converted the Korean measurement system into the metric system, chuck (312.4 mm), chun (31.24 mm), pun (3.124 mm), 1e (0.31 mm).[12]

### Fire-Arrow (Hwa-Jeon)

"The arrow shaft is made of bamboo stick. Its length is 756 mm long, circumference 34.36 mm, the arrow head is made of iron, weight 11 g. Its blades are 14.68 mm wide and 18.74 mm long. Its root is 47 mm long. The tail fins of the arrow are made for feathers, 15.6 mm wide, 194 mm long. The incendiary powder which is 116 mm long, circumference 93.72 mm, is covered with paper or cloth and is covered with oil as a protection against wind and rain. Its is attached to the stem of arrow head and is shot from a bow."[13]

The first Korean rocket was called ju-hwa, which literally means running-fire, and was manufactured between 1377-1389 near the end of the Koryo Dynasty (918-1392) by Choi Mu-Son. The detailed descriptions of running-fire are not available, but indirect information would indicate that it was a rocket-propelled arrow.

The great king Se Jong said to an official of Pyeong-An, Jan-Gil province:

"Running-fire is very efficient and incomparable because it can be fired easily using a quiver by a mounted soldier. It is detrimental to the enemy. Its loud noise and shape instill fear and incite surrender. Once used at night, its exhaust flame lights the fields and shakes the enemy's spirits. When used where the enemy is lying in ambush, its flame and smoke cause the enemy to disclose themselves for fear. Running-fire does not fly straight and it spends more powder and requires more precaution than cannon. . . ."[14]

There are several reasons that point to running-fire being a rocket-propelled arrow.

1. "Running-fire can be fired easily using a quiver..."

    A quiver was not a firearm, it was only an arrow carrier, but running-fire can be fired from a quiver which was made of bamboo, paper or cloth, because the rocket launcher was a simple tube to guide the direction of a launching rocket. Therefore, running-fire had a rocket engine.

2. "Its loud noise and shape..."

    A rocket engine is a jet propulsion device that produces thrust by ejecting combustion gases.[15] Thus, it made loud noises as it ejected combustion gases.

3. "Once used at night, its flame lights the field...its flame and smoke"

    If running-fire were a rifle or a cannon, it could not produce exhaust flames and smoke during flight.

4. "Running-fire does not fly straight"

    Motion of the ancient rockets depended on its stabilizing stick which was made from bamboo and a hole in the powder in the propellant case. Thus, rocket-propelled arrows did not easily fly straight.

5. "Running-fire spends more powder"

    The weight of a solid propellant rocket is, for most part, propellant weight. While the rocket-propelled arrow is propelling, it continues to eject combustion gas. Thus, running-fire used more powder than gun or cannon. Thus, running-fire appears to be a rocket-propelled arrow for the above-given reasons.

## THE KOREAN ROCKET OF THE 15TH CENTURY

The King of the Great Se Jong, who was very concerned about firearms development, was the Fourth King of the Yi Dynasty (1392-1910). Many types of new Korean firearms were developed during the reign of King Se Jong (1418-1450). Development of firearms in the region of King Se Jong is notable as a turning point in the art of making firearms. The Yi Dynasty had stopped imitating the Chinese models and had created a distinctive Korean style.

By the 29th year (1447) of King Se Jong's reign, running-fire was developed into three kinds of rocket-propelled arrows: small-running-fire, medium-running-fire and large-running-fire.

Korea used many running-fire rocket-propelled arrows until 1448. After 1448, it was not seen in *Chon Son Wang Cho Silok* and at the same time, new firearms, magical-machine-arrows, begin to be seen in the same book. According to the *Hwa Po Sick Eon Hae*,

> "The powder tube of the running-fire is equal to the powder tube of the medium-magical-machine-arrow. The tube of the medium-magical-machine-arrow is made from one tenth of a paper which is 2.5 sheets of uncut Korean paper, the length of the powder case is 200 mm, weight of black powder in a powder tube is 44 g."[16]

Dimensions of the medium-magical-machine-arrow of the *Hwa Po Sick Eon Hae* are the same as in the "Firearms Illustrations's". The length of the running-fire powder tube in the *Hwa Po Sick Eon Hae* and length of the medium-magical-machine-arrow's propellant case in the "Firearms Illustration" are the same, which was 200 mm. Therefore, running-fire was replaced with magical-machine-arrow (sin-gi-jeon).

Four kinds of the magical-machine-arrow were constructed: small (so), medium (chung), large (dae) and multiple-bomblets-magical-machine-arrow (san-hwa-sin-gi-jeon).

### Large-Magical-Machine-Arrow (Dae-Sin-Gi-Jeon)

> "The propellant case is made of paper, length 695 mm, external circumference 299.9 mm, thickness 17.8 mm, internal diameter 63.1 mm, the length from the end of the propellant case to the attachment twine is 48.42 mm, the diameter of the hole in the bottom is 37.5 mm. The shaft is made of bamboo stick 5.31 m long, the upper circumference is 31.28 mm, lower circumference is 93.72 mm. The tail fins are made of feathers, 31.28 mm wide, 843.48 mm long. The length from the end of the bamboo stick to the fins is 843.48 mm."[17]

The detailed internal structure of the large-magical-machine-arrow's propellant case and warhead was as follows (Figure 2).

The propellant case was charged with black powder and the top of it was sealed with paper several times. On top of it, a large-magical-machine-arrow-explosive tube (dae-sin-gi-jeon-bal-hwa-tong) was attached and the fuse connects the powder of the propellant case to the large-magical-machine-arrow explosive tube.[18] The magical-machine-arrow-explosive-tube used a cylindrical paper tube filled with black powder with both ends capped. Explosive-tube was divided into 4 groups: large-magical-machine-arrow, large, medium and small-explosive-tube.

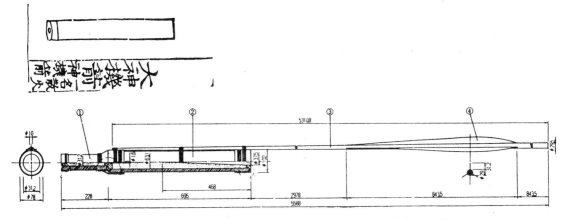

**Figure 2** Drawing of the large-magical-machine-arrow's propellant case in the <u>Kuk Cho Ore Sorye</u> (1474) and new plan of the large-magical-machine-arrow in the <u>Early Firearms in Korea</u> (1377-1600), 1) warhead, 2) propellant case, 3) bamboo shaft, 4) fins.

According the the "Firearms Illustration", the large-magical-machine-arrow explosive tube was as follows:

> "The large-magical-machine-arrow-explosive-tube is 228.1 mm long, 234.3 mm in circumference, 23.1 mm thick, 31.24 mm in internal diameter and is made of paper. Length from the end of the cylindrical case to the attachment twine is 15.62 mm. It has a hold 3.12 mm in diameter at the bottom into which a fuse was inserted."[19]

## Multiple-Bomblets-Magical-Machine-Arrow (San-Hwa-Sin-Gi-Jeon)

> "The multiple-bomblets-magical-machine-arrow has almost the same dimensions as the large-magical-machine-arrow, but the former's warhead was a large-magical-machine-arrow explosive-tube attached on the head of its paper-propellant case, but the multiple-bomblets-magical-machine-arrow's explosive system was in the propellant case.

The detailed internal structure of the multiple-bomblets-magical-machine-arrow's propellant case and explosive system is as follows (Figure 3).

> The lower part of the propellant case was bound with twine as was the large-magical-machine-arrow. The propellant case was charged with powder up to 579.5 mm, leaving 115.59 mm empty. Then several layers of paper were attached to the top surface of the powder. Then several land-fire-tubes (ju-hwa-tong) attached to small-explosive-tubes (so-bal-hwa-tong) were placed into the top of the propellant case with their fuses attached to the propellant charge."[20]

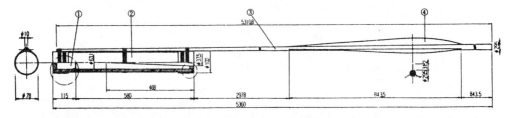

**Figure 3** New Plan of the multiple-bomblets-magical-machine-arrow in the <u>Early Firearms in Korea</u> (1377-1600), 1) warhead, 2) propellant case, 3) bamboo shaft, 4) fins.

## Medium-Magical-Machine-Arrow (Chung-Sin-Gi-Jeon)

"The arrow shaft is made of a bamboo stick. Its length is 14.55 mm, upper circumference is 14.68 mm and lower circumference is 24.99 mm. The arrow head is made of iron, weight 5.5 g. The stem of the arrow-head is 37.4 mm long and 16.56 mm in circumference. Its blades are 13.4 mm wide and 181.2 mm long. The propellant case, 200 mm long, is made of paper, its external circumference is 8.75 mm and its thickness is 6.2 mm, internal diameter is 16.6 mm. The length from the end of the propellant case to the attachment twine is 9.37 mm, the diameter of hole in the bottom is 7.19 mm."[21]

The detailed internal structure of medium-magical-machine-arrow propellant case is as follows (Figures 4 and 5).

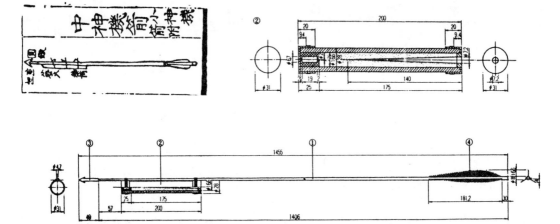

**Figure 4** Drawing of the medium-magical-machine-arrow in the <u>Kuk Cho Ore Sorye</u> (1474) and new plan in the <u>Early Firearms in Korea</u> (1377-1600), 1) arrow shaft, 2) propellant case, 3) arrow head, 4) fins.

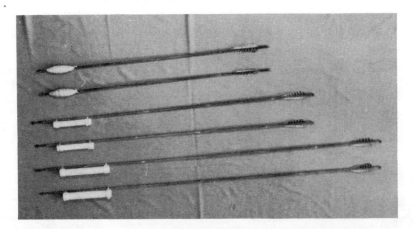

**Figure 5** Fire-arrows, small and medium-magical-machine-arrows (Hang-ju castle memorial museum).

"The lower part of the propellant case was bound with twine as in the large-magical-machine-arrow's propellant case. The propellant case was charged with powder up to 175.09 mm leaving 24.99 mm. A small-explosive-tube (so-bal-hwa-tong), 56.23 mm long,

8

47.8 mm in circumference, 4.37 mm thick and 6.87 mm in internal diameter was inserted. The length from the end of the cylindrical case to the attachment twine was 6.25 mm and the diameter of the nozzle was 2.19 mm. Finally, the powder of small-explosive-tube and the powder of the medium-magical-machine-arrow's propellant case were connected with fuses."[22]

## Small-Magical-Machine-Arrow (So-Sin-Gi-Jeon)

"The arrow shaft is made of bamboo stick 1030.9 mm long. Its upper circumference is 14.68 mm and the lower circumference is 24.99 mm. The arrow head of the small-magical-machine-arrow is the same as that of the medium-magical-machine-arrow. The three tail fins of the arrow are made of feathers, 14.68 mm wide and 146.83 mm long. The 153.08 mm long propellant case is made of paper. Its external circumference is 67.17 mm, 5.0 mm thick, internal diameter 11.56 mm, length from the end of the propellant case to the attachment twine (bound the same way as large-magical-machine-arrow's propellant case) is 9.37 mm. It is charged with powder in the same way as the land-fire-tube, the diameter of the nozzle in the bottom is 4.06 mm. It does not have any explosive tube."[23] (Figure 6)

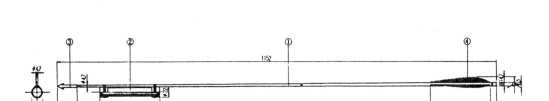

**Figure 6**  New plan of the small-magical-machine-arrow in the <u>Early Firearms in Korea</u> (1377-1600)

## The Manufacturing Method of the Rocket's Propellant Charge

"A long cone-shaped awl was inserted through the hole in the bottom of the propellant case,[24] then a small quantity of powder was spread and hardened with empty cylindrical iron stick. This process was repeated until the case was filled up to a desired level. Then the long awl was taken out leaving the cone-shaped central cavity in the charged propellant case, then it received a fuse.

The external diameter of the cylindrical iron stick was equal to the internal diameter of the propellant case. The internal diameter of the cylindrical iron stick was equal to the diameter of the hole in the propellant case's bottom."[25]

According to the *Wu Pei Chih*, "If the rocket-propelled arrow is to fly straight, the hole must be straight, otherwise it will go off at a tangent",[26] Therefore it was very important to make the rocket engine correctly.

Korean rockets had a paper tube rocket engine which was attached to the top of a bamboo stick serving the function of a stabilizing bar that helped the magical-machine-arrow fly straight. Notably the large Korean rockets had three fins attached to the bamboo stick.[27] The cylindrical paper propellant case had a hole in the bottom believed to be the place for the nozzle where the flame of the burning propellant was ejected for the propulsion of the shell. The Korean rocket's ratio of nozzle diameter to the internal diameter of propellant case was 1:2.3. Large, multiple-bomblets and medium-magical-machine-arrows had a warhead or explosive in front of the propellant case so as to explode over the target area.

Korea had 33,000 running-fires and magical-machine-arrows in 1447,[28] which was a large number for those type weapons, and they were a significant part of the total armaments.

**ROCKET LAUNCH DEVICE**

King Mun Jong was very interested in development of firearms. When he was the Crown Prince, he was one of the persons responsible for the Bureau of Weapons. He invented the fire-cart (hwa-cha) shown in Figure 7, used to launch large numbers of rockets rapidly - also to transport and aim during the battle.

According to the *Mun Jong Silok* (Veritable records of the King Mun Jong era), the fire-cart was invented and tested in February 1451 by King Mun Jong who was the 4th king of Yi Dynasty.[29]

There was a wooden launcher on top of this vehicle, on which were installed 100 medium-magical-machine-arrows or 50 four-arrow-guns.

The detail structure of the fire-cart according to the "Firearms Illustrations" follow.

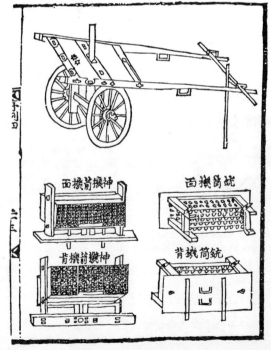

**Figure 7** Drawing of the fire-cart and magical-machine-arrow-launcher in Kuk Cho Ore Sorye (1474).

## Fire Cart

The diameter of the wheel is 874.7 mm. The hub is made of wood, 224.9 mm wide and 206.2 mm in diameter. Each wheel has 15 spokes. The axletree is made of wood, 1312.1 mm long and consists of three parts: a middle square pillar 687.3 mm long and two 312.4 mm long end columns which are inserted into two wheels. Two wide posts, 546.7 mm long, 234.3 mm wide and 62.5 mm thick were set up at both ends of the top side of the middle square pillar. A small post, that has a square bottom pillar of length 231.2 mm and a round upper pillar of length 453.0 mm, is set up at the center of the square pillar. Thin wooden boards are attached between each wide post and the small two-part post. A lower center crossbar, 546.7 mm long, 118.7 mm wide and 46.9 mm thick, which has a hole in the center for the small post, is placed on the two wide posts. Two yokes, 2311.8 mm long, are set up at both ends of the wide posts. It consists of two parts: a four cornered part, 874.7 mm long, 94.7 mm wide, 103.1 mm thick and a column part, 1437.0 mm long and 46.7 mm in diameter. An upper center crossbar, whose length, width and thickness are the same as that of the lower center crossbar is placed between the end of both wide posts. It has a hole in the center for the small post. A rear crossbar whose length and width are the same size as center crossbars, but is 78.1 mm thick. It is attached at the rear (four-cornered part) of both yokes. Then several thin wooden boards are attached between the rear crossbar and the upper center crossbar. This forms a wood box to hide some arms. Its length and width are 624.8 mm by 515.5 mm. Four U-shaped nails are driven into the front and middle of the column part of the yokes to insert rods, which are used to pull the fire-cart.[30]

The cart could be drawn by two men on level ground. It would require another man pushing from behind when going uphill, and two more men had to push the cart when it was going up a steep hill.[31]

New plans for the fire-cart, (Figure 8) were made from the above explanations and an original drawing of the fire-cart in "Firearms Illustration".

It carried a multiple-rockets-launcher (sin-gi-jeon-gi) or a box installed with 50 four-arrow-guns[32] which can shoot 200 thin-arrows[33] at the same time.

## Magical-Machine-Arrow-Launcher (Sin-Gi-Jeon-Gi)

The first crossbar is 1171.55 mm long, 109.3 mm wide, 46.9 mm thick and has a 62.4 mm diameter hole in the center into which the small post of the fire-cart is inserted. Two small square columns which are 171.8 mm long, 62.4 mm wide and thick, are set up on either side of the hole in the crossboard. Two columns, 687.3 mm height, 78.1 mm wide and 31.2 mm thick, are set up near the end of the first crossboard. The spacing of the columns is 843.5 mm. The second crossboard, 843.5 mm long, 109.3 mm wide and 31.2 mm thick, has a hole in the same position as the first crossboard. It is positioned 87.5 mm above the first crossboard. Two detailed boards, 359.3 mm long and 87.5 mm wide, are set up on top of the second crossboard such that the outside ends of both detail-boards are attached to both columns. Two assistant columns, 203.1 mm long, 31.2 mm wide and 15.6 mm thick, are set below the two detailed boards, the spacing of the assistant columns is about 187.4 mm. Both ends of the assistant columns project 31.2 mm below the first crossboard. Four wide boards are attached between the two columns and the second crossboard. This construction is a shallow rectangle box. The top and bottom wide boards have a length of 827.8 mm, a width of 281.2 mm and a thickness of 21.9 mm. The right and left wide boards have a length of 437.4 mm; the width and thickness are the same size as top and bottom. At the center of the bottom wide board there is a hole, which is the same size as the second crossboard's. A rectangle block of wood, 265.5 mm in length, 234.3 mm in width and 56.2 mm in thickness, has the same size hole in it as does the second crossboard. The cylindrical-hole-wood-block is 234.3 mm long, 56 and 22 mm wide and thick has a cylindrical hole, 234.3 mm long, 46.9 mm in diameter bored through it to be loaded a medium-magical-machine or a small-magical-machine-arrow (rocket).

There are 100 cylindrical-hole-wood-blocks in the shallow rectangle box. It consists of seven rows of cylindrical-hole-wood-blocks. The first row has five cylindrical-hole-wood-blocks on both sides of the rectangle block of wood and all of the other rows have 15 cylindrical-hole-wood-blocks. The end of the cylindrical-hole-wood-block is pierced with a wire, then both ends of the wire are attached at right and left wide boards.

Finally, the third crossboard, which measures 843.4 mm in length, 62.4 mm in width and 31.2 mm in thickness, is set up on the top wide boards.[34]

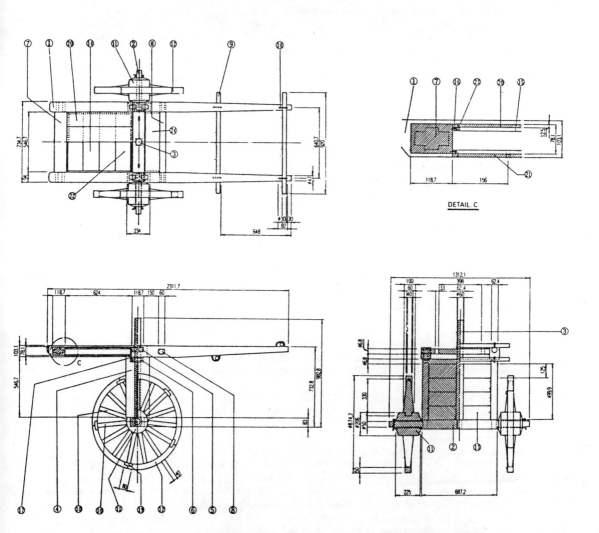

**Figure 8** New plan of the magical-machine-arrow launcher in the Early Firearms in Korea (1377-1600), 1) yoke, 2) axletree, 3) small post, 4) wide post, 5) upper center crossbar, 6) lower center thin wooden board, 15,16,21,23) thin wooden beam, 18) U-shape nail.

The author made a plan (Figure 9) of the magical-machine-arrow-launcher from the above explanations and a drawing (Figure 7) of the magical-machine-arrow-launcher, and constructed a magical-machine-arrow-launcher and fire-cart (Figure 10) from these plans.

The fire-cart with the magical-machine-arrow-launcher was scientifically designed. A cart body that was raised above the axle by short pillars so as to regulate the angle of the cart body from zero to forty-three degrees, so that one can control the rocket's launch angle from zero to forty-three degrees. The magical-machine-arrow-launcher had 100 rocket launching holes. Therefore, it could launch 100 medium or small-magical-machine-arrows in groups of 15 at a time, in quick succession. In peace time, the fire-cart, without its launcher, was used as a simple cart. It was a useful multiple-rocket launcher cart.

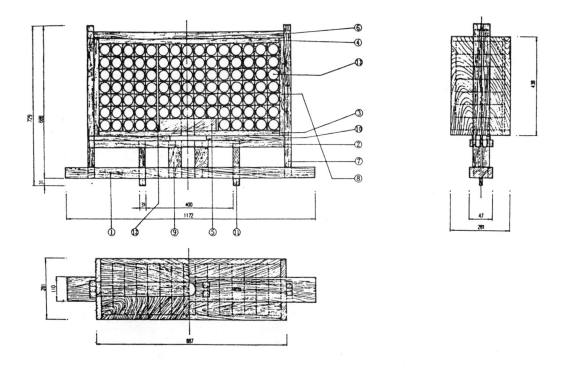

**Figure 9** New plan of the magical-machine arrow launcher in the Early Firearms of Korea (1377-1600), 1) first crossboard, 2) second crossboard, 3-4) top, bottom wide boards, 5) rectangle block of wood, 6) third crossboard, 6) column, 8) side wide board, 9) small square column, 10) detailed board, 11) assistant column, 12) thin column, 13) cylindrical-hole-wood block.

According to the *Mun Jong Silok*, 700 fire-carts were built in Korea in 1451.[35]

Documents on the large-magical-machine-arrow's launcher have not been found, but the large-magical-machine-arrow's total length was 5.7 m; therefore it used a large special launcher.

These weapons are believed to have been used as weapons to fight against the Northern Chinese and Southern Japanese bandits.

**Figure 10** Author's construction of fire-cart and magical-machine-arrow launcher (Hang-ju castle memorial museum).

## CONCLUSIONS

1. The first Korean rocket whose Korean name is "ju-hwa", called running-fire, was used between 1377-1447. In 1448; it was replaced with the magical-machine-arrow (sin-gi-jeon) which was built in four configurations, small, medium, large and multiple-bomblets-magical-arrow. Especially, the large-magical-machine-rocket-propelled-arrow's cylindrical paper case was 70 cm long and 9.5 cm in exterior diameter. It was attached to a 5.3 m bamboo guide stick. The warhead was attached to the head of the propellant case. It was a large paper propellant case rocket.

2. The detailed structure of the four kinds of magical-machine-arrow in Korea after 1448 will furnish information on the early oriental rocket's detailed structure.

3. The fire-cart, multiple-rockets-launcher-cart, was scientifically designed to launch 100 medium or small-magical-machine-arrows in groups of 15 at a time, in quick-succession. The angle of launch was controllable from zero to forty-three degrees.

4. Korea made precision rockets and firearms in the 15th century, because at that time the minimum unit of the Korean measurement system was a "1e" (0.31 mm) for design of rocket, fire-cart and other firearms.

## NOTES AND REFERENCES

1. New plans which were already introduced in the book, *Early Firearms in Korea* (1377-1600), Il-ji Publishing Co., Seoul, 1981, by Chae, Yeon, Seok, were made from the descriptions and drawings in the *Kuk Cho Ore Sorye*. These plans were used to build modern copies of all the firearms used from 1448 to 1451 (4 kinds of rockets, 13 kinds of guns or cannons, 6 kinds of bombs, 13 kinds of projectiles, a rocket launcher cart and an armed cart), and some of those were fired in January 1981. All of these firearms are on permanent display at the Hang-ju castle memorial museum, which is near the city of Seoul in Korea.

2. It was officially compiled by the Board of Rites in 1474 in Korea.

3. Ho, Son-Do, *Rok Sa Hak Po*, Vol. 24, 1964, pp.1-60, Vol. 25, pp.39-98, Vol. 26, 1965, pp.141-165.

4. "The Development of Firearms in Korea 1474-1592", *Rok Sa Hak Po*, Vol. 30, 1966, pp.40-107, Vol. 31, pp. 67-127, "On the Chon-ja Cannon Dated 1555, *Mi Sul Cho Ryo* (National Museum of Korean Art Magazine), Vol. 10, 1965, pp.5-14.

5. Veritable record of the Yi Dynasty, 1413-1865, which was compiled by the veritable record office, Yi Dynasty.

6. "Modern Firearms", in Jeon, Sang-Woon, *Science and Technology in Korea*: Traditional Instruments and Techniques, MIT Press, Massachusetts, 1974, pp.184-206; He seems to have misunderstood the meaning between "arrow" and "rocket". The Korean name "Jeon", literally means the arrow, but he translated it by the English word, rocket. Generally an arrow is not a rocket, but some special arrows were rockets.

7. Boots, T. L., "Korean Weapons and Armor", *Transactions of the Royal Asiatic Society*, Korea Branch, December 1934.

8. Winter, F. H., "The Genesis of the Rocket in China and Its Spread to the East and West," XXXth Congress IAF-79-A-46, p.13 (see *History of Rocketry and Astronautics*, Vol. 10, *AAS History Series*, ed. by Å. Ingemar Skoog, AAS 90-401, pp.3-24, 1990). He mentioned Korean rockets as follows, "Elsewhere in Asia, Montross says the Koreans. But Hagerman, in his detailed study of these engagements, mentions no rockets, only cannon, Partington does not mention Korean rocket weapons..."

9. Wang Lin, "On the Invention and Use of Gunpowder and Firearms in China", *Isis*, Vol. 37, 37, July 1947, p.176.

10. History of Koryo Dynasty, which was officially compiled by Chong, In-Ji et al. in 1451, modern reprints, Yonsei University Press, Korea, 1955.

11. Chae, Yeon-Seok, "A Study of the Korean Fire-Arrows", *Journal of Korean Science History*, 1979, "3. Korean Fire-Arrow", *Early Firearms in Korea (1377-1600)*, pp.22-38, Sun, Fang-Toh, "Rockets and Rocket Propulsion Devices in Ancient China", XXXI Congress, IAF-80-LAA-02, p.9 (see *History of Rocketry and Astronautics*, Vol. 10, *AAS History Series*, ed. Å. Ingemar Skoog, AAS 90-402, pp.25-40, 1990).

12. Park, Heung-su, "A Study of the Korean Measures", *Dai Dong Mun Hwa*, Vol. 4, 1967.

13. *Kuk Cho Ore Sorye*, Vol. 4, p.21.

14. Ho, Son-Do, op. cit. (part 3), *Rok Sa Hak Po*, Vol. 25, p.71.

15. George P. Sutton, Donald M. Ross, *Rocket Propulsion Elements*: An Introduction to the Engine of Rockets, 4th ed., Wiley Sons, New York, 1976, p.1.

16. Lee So, *Hwa Po Sick Eon Hae*, 1635, p.32.

17. *Kuk Cho Ore Sorye*, Vol. 4, pp.20-21.

18. There is an "explosive tube" in the description of the large-magical-machine-arrow but the large-magical machine-arrow-explosive-tube was among the kinds of explosive-tubes in the "Firearms Illustration". Therefore the explosive-tube of the large-magical-machine-arrow meant a large-magical-machine-arrow-explosive-tube.

19. *Kuk Cho Ore Sorye*, Vol. 4, pp.17-18.

20. Ibid., pp.20-21.

21. Ibid.

22. Ibid.

23. Ibid.

24. Tenney L. Davis & James R. Ware, "Early Chinese Military Pyrotechnics", *J. of Chemical Education*, Vol. 24, November 1974, p.532. He called it "the central cavity in the Rocket propelling charge" in his study.

25. *Kuk Cho Ore Sorye*, Vol. p.17.

26. Tenney L. Davis, op. cit., p.532.

27. Other large rocket-propelled arrows did not use fins on stabilizing bar, Å. Ingemar Skoog, "The Swedish Rocket Corps 1833-1845", *Essays on the History of Rocketry and Astronautics,* NASA CP-2014, Vol. 1, p. 10-20, Fig. 2, 12 (see *History of Rocketry and Astronautics*, Vol. 7, Part I, *AAS History Series*, ed. by R. Cargill Hall, AAS 86-501, pp.9-22). F. W. Foster Gleason, "Lost Causes" *Gun Digest*, ed. John T. Amber, pp. 36-39. Winter, F. H., "On the Origin of Rockets", *Chemistry*, Vol. 49, No. 2, p.19.

28. Chae, Yeon-Seok, *Early Firearms in Korea (1377-1600)*, p.69.

29. Ho, Son-Do, "The Introduction and Development of Firearms in Korea 1356-1474", part 2, p.52.

30. *Kuk Cho Ore Sorye*, pp.22-24.

31. Jeon, San-Won, *Science and Technology*, p.199.

32. It was one of the Korean rifles in 1448. It was able to shoot 4 thin-arrows (bullets) at the same time. Its barrel length was 180.1 mm and the diameter of the muzzle was 21.9 mm. Therefore, it was called a four-arrow-gun.

33. It was one of the bullets in Korea in 1448. It was like a small arrow. Its total length was 218.6 mm.

34. *Kuk Cho Ore Sorye*, Vol. 4, pp.25-26.

35. Chae, Yeon-Seok, *Early Firearms in Korea (1377-1600)*, p.169.

# Part II

# ROCKETRY AND ASTRONAUTICS: CONCEPTS, THEORIES, AND ANALYSES

## Chapter 2

## LEONHARD EULER'S IMPORTANCE FOR AEROSPACE SCIENCES - ON THE OCCASION OF THE BICENTENARY OF HIS DEATH[*]

### Werner Schulz[†]

Two hundred years ago, on 18 September 1783, Leonhard Euler, aged 76, died in St. Petersburg. He was the most important mathematician of his time, and one of the greatest mathematicians of all times. In view of his extraordinary creativity, he made outstanding contributions in almost all fields of mathematics, in mechanics and their related fields of application, as well as in physics and in the technical sciences. Algebra, theory of numbers, calculus, theory of differential equations, theory of functions, calculus of variations, geometry, topology, celestial mechanics, fluid dynamics, ballistics, optics, naval sciences, and engineering owe basic findings to him.

Euler's productivity was of enormous extent. When in 1907, on the occasion of Euler's 200th anniversary, the Schweizerische Naturforschende Gesellschaft decided to start, with the help of foreign scientific institutions, on a complete edition of Euler's works; its scope was approximated at 45 volumes. It was planned to have three series: I. Opera mathematica; II. Opera mechanica et astronomica; III. Opera physica, Miscellanea. In 1910-13, the Swedish historian of mathematics, Gustaf Eneström, compiled an index of Euler's works amounting to 866 references.[1] The number of volumes first envisaged soon proved to be insufficient, and in the meantime, it has become apparent that, including a Series IV containing Euler's correspondence and his manuscripts, the number of volumes will be roughly twice as many as assumed originally.[2] The first volume was published 1911. Series I has by now been completed. Concerning the further series, some volumes are still under preparation.

Apart from having written a great number of papers, the majority of which appeared in the publications of the Academies of Sciences of St. Petersburg and Berlin, Euler is the author of more than a dozen compendia, of several hundred pages each in most cases, which have influenced the development of mathematics

---

[*] Presented at the Seventeenth History Symposium of the International Academy of Astronautics, Budapest, Hungary, 1983.

[†] Deutsche Gesellschaft für Luft- und Raumfahrt (DGLR), Braunschweig, Federal Republic of Germany. Dr. Schulz died in 1984.

and mechanics far beyond his time.[3-15] Many of his findings also bear on the solution of problems in the field of aerospace sciences.

Leonhard Euler (1707-1783)

Apart from the diversity of the fields of research covered and the amount of publications, the range of ideas in Euler's papers calls for admiration. His treatises are of great lucidity, enabling his reader to follow his trains of thought very closely.

## EULER'S LIFE AND WORK

Leonhard Euler, born 15 April 1707 in Basel as a parson's son, started out studying theology in his home town, but very soon and with his father's consent changed over to mathematics. His teacher was the well-known Johann Bernoulli (1667-1748), and when in 1725 Bernoulli's sons Nikolaus (1696-1726) and Daniel (1700-1782) who were very close friends of Euler's were appointed to the Academy of Sciences in St. Petersburg, recently founded by the Empress Catherine I, it was Euler's wish to follow them as soon as possible. This ambition was realized in 1727, when Euler had reached the age of 20, but the timing was inauspicious: Euler arrived in St. Petersburg on the day of Catherine's death, and because of the political insecurity, the future of the Academy was uncertain as well. Euler, therefore, had to resign himself, for the time being, to an appointment with the Russian navy. In 1730, he obtained a professorship in physics at the Academy, and in 1733, when Daniel Bernoulli returned to Basel, he was appointed as his successor for the chair of mathematics. After being thus established, he married in the same year.

Euler's first mathematical treatises were influenced by the Bernoullis. In particular, two investigations on geodetic curves, written in 1728/29[16,17] and a paper on isoperimetric problems[18] caused a stir; they became important for the development of the calculus of variations. Investigations on the theory of numbers, however, also made the scientific community note the then 25 year old Euler.

It is important that during this time Euler concerned himself intensively with mechanics. He undertook to write a compendium on mechanics[3], in which he systematically developed each individual topic. The first volume was completed in 1734, the second one followed in 1736. The innovation this work presented was the consequent application of analytical methods drawing on Leibniz' form of calculus rather than on the more circumstantial geometrical-synthetic procedures used hitherto. For a systematic progression of this kind, Euler had to fill in many gaps through contributions of his own. The work had an enormous impact, and its influence on the way of thinking of mathematicians, physical scientists and engineers has continued until today.

In 1735 Euler had the misfortune to lose his right eye because of an abscess; his productivity, however, remained unbroken. In 1740, following the death of Empress Anna, during whose ten year reign the Academy had been promoted extensively, a new period of uncertainty started. In the same year, Frederic II became King of Prussia, and one of the first tasks he concerned himself with was the reorganizing of the Berlin Academy of Sciences. Through d'Alembert, who was his advisor in these matters, his attention was drawn to Euler as the most important mathematician. Thus it happened that in 1741 Euler moved to Berlin where he worked for 25 years.

In 1744, Euler published two further compendia of major importance. Having concerned himself from the very beginning with problems of the calculus of variations, and having gradually developed the necessary means for the solution of these problems, he now presented the first extensive survey of this branch of mathematics.[4] What distinguishes this work is the fact that Euler gives approximately 100 examples for application in order to illustrate the methods.

The second compendium which appeared in 1744 deals with the theory of planetary and cometary motions.[5] It remained of importance until well into the 19th Century, in particular because of its dealings with disturbed planetary orbits, and some of its findings are still of interest today.

In the 1740s and 1750s, Euler was also confronted with technical problems in Berlin. They concerned ballistics and hydraulic machinery.

When Frederic II approached him to learn which was to be considered the best work on practical questions of artillery, Euler named for him the book of the Englishman Benjamin Robins, *New Principles of Gunnery*, which had appeared in 1742, and volunteered to translate it into German.[33] The strong point of Robins' work is that it is based on numerous shooting experiments and exact measurements. There were, however, major deficiencies as far as the theory of ballistics was concerned, because Robins had not been able to keep abreast of the developments in mathematics and mechanics. Through the comments and annotations which Euler added to his translation the first extensive outline of interior and exterior ballistics drawing on calculus came into existence. The impact of Euler's edition of Robins' book, which appeared in 1745,[6] was such that it was retranslated into English, and, upon the request of the French Minister for Naval Affairs and Finance, it was also translated into French. In France, it was introduced as a basic reader in the schools of artillery.

The investigations on hydraulic machinery, which resulted in several publications,[19-21] were stimulated by the waterwheel suggested by Andreas Segner (1704-1777), a professor at the University of Göttingen. Euler calculated the acceleration of the water, defined the turning moment the water causes to the wheel, and indicated the resulting pressures also in non-steady cases. In particular, it is worth mentioning that by analyzing the losses which occur with the Segner waterwheel, Euler arrived at the suggestion of introducing guide vanes in order to obtain a well defined inlet flow and thus reduce the losses. He realized the importance of the problem of cavitation: The use of blades was to ensure a continuous transfer of the water from stator to impeller. It is interesting to note that the Swiss firm of turbine construction, Escher-Wyss, built a model turbine according to Euler's suggestions in 1943 which had a maximum efficiency of 71%--a noteworthy result indeed.[34]

Within the field of pure mathematics, Euler compiled during his time in Berlin three major compendia, which remained the leading textbooks for a whole century: *An Introduction into the Analysis of the Infinite* in two volumes (published 1748), a book on differential calculus with examples (1755)[8] and three volumes on integral calculus dealing also with ordinary and partial differential equations as well as with the calculus of variations (completed in 1763, but published only after Euler had already left Berlin in 1768 for St. Petersburg).[12-14] In these books numerous terms which are customary today were first introduced.

Euler also concerned himself with problems of the theory of cognition. In 1750, he published a paper on space and time, which had its influence on the philosopher Immanuel Kant (1724-1804).[22]

From 1744 onwards, Euler was director of the mathematical section of the Berlin Academy of Sciences. When the president of the Academy, Maupertuis (1698-1759), was not available, Euler acted as his official deputy. He was not on very good terms with Frederic II, however. Their personalities were too different. In comparison to the French, who were highly regarded by the German King, Euler lacked the attitude of the man-of-the-world and his religious and conservative ideological outlooks were incompatible with those of Frederic II, who himself had no deep understanding of mathematics. Thus, after the death of Maupertuis, Euler did not become president of the Academy, as one should have expected in view of his outstanding academic contributions. Instead, the King continued to rely on the guidance of d'Alembert from Paris concerning the running of the Academy.

While in Berlin, Euler remained in close contact with the Academy in St. Petersburg and also published papers there. When, in 1762, after Catherine II had taken over, St. Petersburg attempted to obtain his services again, these efforts finally led Frederic II to agree in 1766 for Euler to leave Berlin.

Shortly after the beginning of Euler's second period in St. Petersburg, the eyesight of Euler's remaining left eye was reduced because of a cataract. A famous specialist on cataracts operated on him in 1771, but without success. Thereafter, Euler was almost completely blind. His creativity, however, remained unbroken. His students and assistants - foremost among them his own eldest son Johann Albrecht (1734-1800), himself a reputed mathematician, as well as Nicolaus Fuss (1755-1825), who had been recommended to Euler as a very promising young man from Basel, and the astronomer Anders Johan Lexell (1740-1784), who came from Finland originally - had to read to him, and he dictated to them. Sitting at a desk with a slate top, Euler himself was able to write out formulas in big letters, and since he furthermore enjoyed fantastic powers of memory, their resulted not only numerous further papers on a variety of topics from mathematics, mechanics and physics, but also a number of copious books, among them a theory of the motions of the Moon. After having already written a treatise on this difficult subject in Berlin in 1753,[9] in 1772 a second book followed, containing astronomical tables which were of great importance for navigation at sea.[15] Twice during this time, Euler also took part in competitions within this field for which the Paris Academy had invited entries.[23]

Euler died of a stroke in 1783, in the middle of his work. On his sheet of slate there were formulas he had been working on. They concerned a recent event. In the summer of 1783; the brothers Montgolfier had been successful with the first ascent of a hot air balloon. When Euler learned about this, he engaged himself in approximately calculating the ascent of such a balloon under realistic presumptions. Later, his son turned over his calculations to the Paris Academy of Sciences for publication.[24]

## FINDINGS FROM EULER'S WORKS WHICH BEAR ON AEROSPACE SCIENCES

We come across Euler's name in connection with formulas and theorems in the most varied fields, beginning with the Eulerian straight line of the triangle of elementary geometry and the polyhedron theorem on the relation between the numbers of the vertices, faces and edges of a polyhedron of topology, for which Euler was the first to supply a strong proof, up to the Eulerian sum formula of the theory of series, from which the Eulerian constant can be derived, and Euler's integral of a second kind for the gamma function. Hereafter, some of the findings from Euler's works shall be discussed which are of importance for aerospace sciences.

It was a great merit of Euler's works that they contributed considerably to spread in the use of Leibniz' form of calculus. Furthermore, Euler helped to introduce terms which are now customarily used in the sense of Euler by everybody who concerns himself with mathematics, whereas hitherto there was no uniformity. This applies for the terms f(x) as functional symbol and $\sum$ as summation sign, for the symbols $e$ for the base of the natural logarithm, $\pi$ for the circular number, and $i$ for the imaginary unit (with the relation between these three numbers $e^{i\pi} = -1$), as well as $sin$ and $cos$ for the trigonometric functions. Euler also indicated the connection between the exponential function and the trigonometric functions:

$$e^{ix} = \cos x + i \sin x.$$

In flight mechanics and guidance and control, the three Eulerian angles are used in order to describe the position of a body. Euler proved that they are sufficient for defining the inclinations of three body-fixed axes against three space-fixed axes. Of major importance for guidance and control are furthermore the Eulerian gyroscopic equations.

With regard to general mechanics, Euler's explanations of the momentum and angular momentum theorems with the development of a method for determining the motion of rigid bodies are most famous and of fundamental importance.[25,26]

Euler' succeeded in fluid mechanics in supplying the equations of motion for frictionless fluids. Prior to him, Johann Bernoulli had applied the general laws of mechanics to the one-dimensional movement of fluids. By precisely taking into account the hydrodynamic pressure, Euler dealt with three-dimensional flows. He thus completed the outline of classical hydrodynamics. The fundamental equations he derived still present the starting point for the study of fluid mechanics.[27] Incidentally, attention ought to be paid to the fact that Euler succeeded seven years prior to d'Alembert in deriving the so-called d'Alembert paradoxon, according to which a body in a frictionless flow meets with no resistance. Euler's proof of this theorem is contained in one of the annotations to his translation on Robins' *New Principles of Gunnery.*[35]

Euler also applied the general equations of motion of hydrodynamics to problems of gasdynamics by drawing on them for the investigation of acoustical problems such as the spreading of sound.

Researching the field of ballistics, Euler concerned himself with air drag. In a paper presented to the Berlin Academy in 1752 and published in 1755,[28] he assumed the air drag as proportional to the square of the velocity. He applied a method to the integration of the ballistic equation which served as the basis for exterior ballistics until the 20th Century: Since the differential equation cannot be integrated in closed form, he divided the trajectory into sections for which a quantity variable with respect to time is substituted by a constant average, thus arriving at the integration. Regarding the dependence of air drag on the angle of attack $\alpha$ of the flow exposed body, Euler realized that Newton's assumption of the proportionality to $\sin^2 \alpha$ could not be correct.

In applied mathematics, Euler indicated a procedure for the approximate numerical integration of ordinary differential equations of the first order, which represents the starting point for the further developments that eventually led to the well-known Runge-Kutta procedure.

In order to appreciate Euler's contributions to the development of calculus of variations - the name of which was introduced by him for this branch of mathematics - one would have to enumerate almost all major achievements in this field, beginning from the derivation of the so-called Eulerian differential equation as a necessary condition for making an integral to an extremum and the Euler-Lagrange method of multiplicator. However, the methods for the calculus of variations which Euler supplied have also provided the basis for the development of theories for the solution of problems of optimization in our time where the classical prerequisites are no longer given. The older as well as the more recent procedures are indispensable mathematical tools when it comes to finding optimal solutions under certain specific conditions for problems in aerospace sciences.

Out of the numerous examples Euler supplied in his works in order to demonstrate the methods of the calculus of variations came a particularly famous result, i.e. the Eulerian column buckling equation, which plays such an important role in structural mechanics. It is contained in Euler's book on the solution of isoperimetric problems that appeared in 1744.[4] This book has an annex entitled "On Elastic Curves", and there Euler describes how the form of elastic curves can be determined by applying the methods he had just developed. The investigation of the bending of an articulated rod of constant cross-section requires minimizing the integral $1/R^2 \, ds$, $ds$ signifying the element of arc, and $R$ the radius of curvature. By introducing rectangular coordinates, Euler arrived at the differential equation of the elastic curve and he could thus indicate the force requisite in order to bend out the rod by an infinitely small amount. The result is the well-known formula for the Eulerian buckling load.

Euler's interest in investigations on celestial mechanics dates back well into his youth: His diary of 1727 already contained a comment on the general three-body problem. His famous works in this field, in particular his books on planetary and cometary motions[5] as well as on the movement of the Moon[9,15] derive from his time in Berlin and from his second period in St. Petersburg. As with fluid mechanics and the calculus of variations, Euler's works within the field of celestial mechanics became the starting point for those who followed after him and are still of relevance for flight sciences today. This applies in particular to the methods applied and the results arrived at in his investigations on the determination of orbits, on disturbed Keplerian orbits, and on the three-body problem.

Subsequently, those works of Euler's on celestial mechanics shall be dealt with which became important for space flight mechanics. In 1766/67, Euler published treatises on special cases of the three-body problem that allow for analytical solutions. It is the rectilinear motion of three bodies attracting each other according to the Newtonian gravitational law[29] and of the problem of the two fixed centers.[30-32]

Investigating the three-body problem on the straight line,[29] Euler also went into the question of the collision of two bodies whereby the velocities tend to infinity if the distance between the two bodies becomes zero. Euler solves the problem through a transformation of the variable, thus providing the first example for a regularization of the equations of motion, a method which is of major importance for flight mechanics today.

Investigating the movement of a body under the attracting forces of two fixed centers - in (Ref. 30) for the case in which the body moves in a plane, in (Ref. 31) for the case in which its trajectory is a three-dimensional curve - presents the advantage that there appear no centrifugal and Coriolis forces. In celestial mechanics, the posing of this problem, e.g., in the case of the movement of the Moon subjected to the influence of the Earth and the Sun, is only of academic interest, because it is inadmissible to suppose fixed centers: During the lunar cycle the Earth moves significantly around the Sun, approximately $30^\circ$. For the motion of an artificial satellite, however, orbiting around the Earth within a short period, it is quite feasible to suppose the geocentric position of the Moon or the Sun for a certain time as fixed, thus resulting in the Euler case of fixed centers for the three-body problems Earth-Moon-satellite or Earth-Sun-satellite. For solving the problem, Euler introduced elliptic coordinates and a fictitious time. One is led to elliptic integrals. The two-center problem can be used as the starting point for the planar restricted three-body problem whereby two bodies with finite masses turn around their common mass center according to the Keplerian laws. To be found is the trajectory of a third body of negligible mass moving in the same plane under the influence of the first two bodies. By applying a coordinate system rotating with the two bodies of finite masses, one is able to reduce the restricted three-body problem to the problem of the two fixed centers. On this basis, numerical calculations for the Earth-Moon trajectories of space vehicles have been made.

In conclusion, mention must be of Euler's last investigation in which he was engaged at the time of his death, concerning the theory of the Montgolfier balloon.[24] Not restricting himself to inquiring into the ceiling, Euler dealt with the problem of dynamics, deriving the differential equation for the ascent of the balloon, which he transformed by drawing on the energy theorem, thus arriving at a linear differential equation of first order which enabled him to say something about the speed of ascent and the length of path of the balloon in dependence on time. His assumptions for the calculations were as follows: The balloon was a sphere of constant volume and constant weight. Air temperature was to be constant, i.e., an isothermic atmosphere is to assumed. The air drag was assumed to be proportional to the air density, diminishing exponentially with the height, as well as proportional to the square of the speed. The drag coefficient was assumed to be constant, equal to 0.5. Under these assumptions the differential equation for the ascent is integrated approximately. In 1945, J. Ackeret engaged in an in-depth study of Euler's investigation, thereby discovering a fault in the calculation of the ceiling, which

Euler himself would undoubtedly have realized had he lived to see the paper published.[36] Ackeret's conclusion was:

> "Even the final and of necessity incomplete investigation of Euler's shows all criteria which characterize his studies in mechanics: a general conception of dynamics, the definition of a specific mathematical problem, and intelligent simplifications in the process of calculation in accordance with the (then given) possibilities of analysis."

## REFERENCES

1. G. Eneström, Verzeichnis der Schriften Leonhard Eulers. Jahresbericht der Deutschen Mathematiker-Vereinigung, Ergänzungsband IV, 388pp. B. G. Teubner, Leipzig (1910/13).

2. Leonhardi Euleri Opera onmia. Series prima: Opera mathematica (in 29 volumes, 30 volume-parts).-Series secunda: Opera mechanica et astronomica (in 31 volumes, 32 volume-parts).-Series tertia: Opera physica, Miscellanea (in 12 volumes).-Series quarta: A: Commercium opistolicum (in 8 volumes); B: Manuscripta (in approximately 7 volumes). B. G. Teubner, Leipzig/Berlin; Orell Füssli, Zürich/Lausanne; Birkhäuser, Basel/Boston/Stuttgart (1911 ff.). In the following cited as EOo.

**3-15 Books by L. Euler.**

3. Mechanica sive motus scientia analytice exposita, I, II, 496 and 508 pp. Petersburg (1736).-EOo II, 1 and 2 (1912).

4. Methodus inveniendi lineas curvas maximi minimive proprietate gaudentes, sive solutio problematis isoperimetrici latissimo sensu accepti, 324pp. Lausanne/Genf (1744).-EOo I, 24 (1952).

5. Theoria motuum planetarum et cometarum, 187pp. Berlin (1744).-EOo II, 28, 105-251 (1959).

6. Neue Grundsätze der Artillerie. 736pp. Berlin (1745).-EOo II, 14, 1-409 (1922).

7. Introductio in analysin infinitorum, I, II. 321 and 402pp. Lausanne (1748).-EOo I, 8 (1922) and 9 (1945).

8. Scientia navalis, I, II. 491 and 536pp. Petersburg (1749).- EOo II, 18 (1967) and 19 (1972).

9. Theoria motus lunae exhibens omnes ejus inaequalitates, 355pp. Berlin (1753).-EOo II, 23, 64-336 (1969).

10. Institutiones calculi integralis, 904pp. Berlin (1755).-EOo I, 10 (1913).

11. Theoria motus corporum solidorum seu rigidorum, 552pp. Rostock/Greifswald (1756).-EOo II, 3 (1948) and 4 (1950).

12. Institutiones calculi integralis, I. 546pp. Petersburg (1768).-EOo I, 11 (1913).

13. Institutiones calculi integralis, II. 534pp. Petersburg (1769).-EOo I, 12 (1914).

14. Institutiones calculi integralis, III. 647pp. Petersburg (1770).-EOo I, 13 (1914).

15. Theoria motuum lunae, nova methodo pertractata, 791pp. Petersburg (1772).-EOo II, 22 (1958).

**16-32 Treatises by L. Euler**

16. Solutio problematis de invenienda curva, quam format lamina utcunque elastica in singulis punctis a potentiis quibuscunque sollicitata. Comment. Acad. Sci. Petropol. 3, 1728, 70-84 (1732).-EOo II, 10, 1-16 (1947).

17. De linea brevissima in superficie quacunque duo quaelibet puncta jugente. Comment. Acad. Sci. Petropol. 3, 1728, 110-124 (1732).-EOo I, 25, 1-12 (1952).

18. Problematis isoperimetrici in latissimo sensu accepti solutio generalis. Comment. Acad. Sci. Petropol. 6, 1732/33, 123-155 (1738).-EOo I, 25, 13-40 (1952).

19. Recherches sur l'effet d'une machine hydraulique proposée par Mr. Segner professeur à Göttingue. Mém. Acad. Sci. Berlin 6, 1750, 311-354 (1752).-EOo II, 15, 1-39 (1957).

20. Application de la machine hydraulique de M. Segner à toutes sortes d'ouvrages et de ses avantages sur les autres machines hydrauliques dont on se sert ordinairement. Mém. Acad. Sci. Berlin 7, 1751, 271-304 (1753).-EOo II, 15, 105-133 (1957).

21. Théorie plus complète des machines qui sont mises en mouvement par la réaction de l'eau. Mém. Acad. Sci. Berlin 10, 1754, 227-295 (1756).-EOo II, 15, 157-218 (1957).

22. Réflexions sur l'espace et le temps. Mém. Acad. Sci. Berlin 4, 1748, 324-333 (1750).-EOo III, 2, 367-383 (1942).

23. Résponse à la question proposée par l'Académie Royale des Sciences de Paris, pour l'année 1770. Perfectionner les méthodes sur lesquelles est fondée la théorie de la lune, de fixer par ce moyen celles des équations de ce satellite, qui sont encore incertaines, et d'examiner en particulier si l'on peut rendre raison, par cette théorie de l'équation séculaire du mouvement de la lune. Recueil des pièces qui ont remporté de l'Académie Royale des Sciences 9, 94pp. (1777).- Dto. Réponse à la question proposée par l'Académie Royale des Sciences de Paris, pour l'année 1772. Recueil 9, 38pp. (1777).- To be reprinted in EOo II, 24.

24. Calculs sur les ballons aérostatiques faits par feu M. Léonhard Euler, tels qu'on les a trouvés sur son ardoise, après sa mort arrivée le 7 Septembre 1783. Mém. Acad. Sci, Paris 1781, 264-268 (1784).-EOo II, 16, 165-169 (1979).

25. Découverte d'un nouvea principe de mécanique. Mém. Acad. Sci. Berlin 6, 1750, 185-217 (1752).-EOo II, 5, 81-108 (1957).

26. Nova methodus motum corporum rigidorum determinandi. Novi Comment. Acad. Sci. Petropol. 20, 1775, 208-238 (1776).-EOo II, 9, 99-125 (1968).

27. Principes généraux du mouvement des fluides. Mém. Acad. Sci. Berlin 11, 1755, 274-315 (1757).-EOo II, 12, 54-91 (1954).

28. Recherches sur la véritable courbe que décrivent les corps jettés dans l'air ou dans un autre fluide quelconque. Mém. Acad. Sci. Berlin 9, 1753, 321-352 (1755).-EOo II, 14, 413-447 (1922).

29. De motu rectilineo trium corporum se mutuo attrahentium. Novi Comment. Acad. Sci. Petropol. 11, 1765, 144-151 (1767).-EOo II, 25, 281-289 (1960).

30. De motu corporis ad duo centra virium fixa attracti. Novi Comment. Acad. Sci. Petropol. 10, 1764, 207-242 (1766).-EOo II, 6, 209-246 (1957).

31. De motu corporis ad duo centra virium fixa attracti. Novi Comment. Acad. Sci. Petropol. 11, 1765, 152-184 (1767).-EOo II, 6, 247-273 (1957).

32. Problème. Un corps étant attiré en raison réciproque quarrée des distances vers deux points fixes donnés, trouver le cas où la courbe dérite par ce corps sera algébrique. Mém. Acad. Sci. Berlin 16, 1760, 228-249 (1767).-EOo II, 6, 274-293 (1957).

**33-36 Further quotations**

33. B. Robins, New Principles of Gunnery. London (1742).-Reprinted in: J. Wilson, Mathematical Tracts of the Late Benjamin Robins, Esq., Vol. I, 1-153. London (1761).

34. J. Ackeret, Untersuchung einer nach den Euler'schen Vorschlägen (1754) gebauten Wasserturbine. Schweiz. Bauzeitung 123, 2-4 (1944).-See also EOo II, 15, XLII-LI (1957).

35. I. Szabó, Geshichte der mechanischen Prinzipien und ihrer wichtigsten Anwendungen, 242-245. Birkhäuser Verlag, Basel/Stuttgart (1976).

36. J. Ackeret, Leonhard Eulers letzte Arbeit. In: Festschrift zum 60. Geburtstag von Prof. Dr. Andreas Speiser, 160-168. Orell Füssli Verlag, Zürich (1945).

# Part III

# THE DEVELOPMENT OF LIQUID- AND SOLID-PROPELLANT ROCKETS, 1880-1945

# Chapter 3

# THE FOUNDING OF THE JET PROPULSION RESEARCH INSTITUTE AND THE MAIN FIELDS OF ITS ACTIVITY[*]

## B. V. Rauschenbach[†]

Now that astronautics has become an important sphere of man's activity, a large industry comparable in scope with aircraft production has firmly established itself in our everyday life (television, weather forecasting, etc.), and one is tempted to turn to its sources and retrace its history. This year gives us a good opportunity to indulge in such retrospections as it is the 50th anniversary of the founding of the Jet Propulsion Research Institute, an organization specially concerned with jet technology. This institute was the first of its kind, not only in the Soviet Union but in the whole world.

It would not be correct to presume that the institute started from scratch. Long before its emergence, rocket technology enthusiasts in the Soviet Union joined their efforts in the Gas Dynamics Laboratory (GDL) in Leningrad and in the Group for the Study of Jet Propulsion (GIRD) in Moscow.

For in those times, these two organizations rated among the larger ones. The Leningrad GDL mainly specialized in artillery. It was headed first by Tikhonmirov, then by Petropavlovsky. The work they had started led to the emergence of widely known rocket artillery that played an extremely important role during World War II. But rocket artillery was not the only field of the GDL's activity. Suffice it to say that one of its members was Glushko, who devoted himself to the development of liquid-propellant rocket engines. Space orientation, so to speak, was also the trend in a group of Moscow enthusiasts incorporated in the GIRD. Here, the master minds were Tsander and Korolev who investigated problems involving today's astronautics. Since these two organizations worked simultaneously on closely related problems in two different cities, and their leaders, as well as the rank-and-file members, felt the need for joining their not numerous forces in a single organization, the idea naturally arose to create a single jet propulsion research institute. The leading role in the formation of the Institute belongs to the Leningraders. In 1932, Rynin, Perelman, and Petropavlovsky wrote a letter on behalf of the Leningrad scientists in which they proposed to join their efforts. In 1933 similar ideas were put

---

[*]  Presented at the Seventeenth History Symposium of the International Academy of Astronautics, Budapest, Hungary, 1983.

[†]  U.S.S.R. Academy of Sciences, U.S.S.R.

forward by both the Leningrad GDL and Moscow GIRD members, which shows that both organizations strove for a merger.

In order to give investigations in rocketry the necessary scope and raise them, in terms of planning and execution, to the level of state importance, a decision was taken to conduct them within the framework of the People's Commissariat of Heavy Industry, which included at that time both aircraft and artillery plants and institutes. This provided the necessary conditions for establishing business cooperation with relevant bodies and enabled the new organization to rely on the entire might of the aircraft and artillery industries, thus opening much broader possibilities for fruitful activities. In October 1933 the Council of Labor and Defense issued a decree that set up the Jet Propulsion Research Institute (RNII), thus starting the history of this remarkable organization.

The plan of work drawn up right after the formation of the RNII devoted much attention to the development of rocket artillery. It was an important task, and owing to its successful fulfillment, the Soviet Union entered World War II armed with rocket projectiles that could be fired from aircraft and highly mobile (truck-mounted) launchers. The history of the development of rocket artillery armament could well be the subject of a separate report, and goes beyond the scope of this communication, which is confined to the institute's activity in the field of space technology.

As regards the investigations which paved the way for space flight, the main problem addressed was the development of a reliable liquid-propellant rocket engine. Understandably, there was no point in speaking about rockets until such an engine was created. Given priority attention, large-scale work on this problem got under way. Already in 1933 Glushko's engine ORM-52 was subjected to an official trial. The term "official" deserves special explanation. As is known, trials may be of different types. Unlike a routine trial which is conducted by the engine designer himself with a view to obtaining the data he needs, an official trial is aimed at testing an engine against the technical requirements established by an outside organization. During official trials, all formal and technical rules must be strictly observed and all those who have had experience with presenting some products for official trials know only too well the difference between such trials and conventional bench tests arranged for the designer's own needs.

The year 1936 witnessed the official trial of the ORM-65 engine, which was installed in Korolev's rocket 212. It had been also used originally with the rocket glider 318. To make a long story short, it was in small-scale batch production. Mention should also be made of the well-known BI-1 fighter plane flown in 1942. It was powered by a liquid-propellant rocket engine also developed in the Jet Propulsion Research Institute. In 1944 a test was applied to yet another engine designed by Isayev, who was at that time working at the Institute. As we shall see, all the main engines developed in those years for rocket-powered vehicles were directly connected with the activity of the RNII.

A more detailed analysis of the investigations connected with the development of engines reveals the remarkable scope and depth of the institute's scientific endeavor. Of special significance was the intensive researches carried out simultaneously in the possibilities of low-boiling and high-boiling oxidants. The investigations conducted in the first field were aimed at developing alcohol-oxygen engines (with a prospect of changing over to kerosene-oxygen propellants), whereas efforts made in the second field concentrated on the development of engines using kerosene and nitric acid. In both fields, intensive experimental and theoretical work was under way on such problems as the atomization of fuel components by injectors, formation of combustible mixture in engine chambers, as well as the problem of engine cooling, particularly with propellant components (the possibility of such cooling was seriously called in question in those days). Efforts were also made to improve the engine starting procedure (including the procedure of multiple starts), to develop methods for increasing the efflux velocity or, using the terminology of that period, the engine specific impulse, and to develop gas generators without which the modern rocket engine would be simply inconceivable. Among the subjects of scientific investigation were also a few "exotic" themes which died a natural death, such as the use of ceramic lining for protection of the combustion chamber inner surfaces from high temperatures.

As we see, the subjects that were in the focus of the researchers' attention in that period differ but little from those under investigation at present. The only new problem to emerge since that time is that of stability of the engine working process. In the early period of the institute's activity the researchers did not go into the intricacies of stable operation of liquid-propellant rocket engines, and did not study the phenomena of high-frequency oscillations. As regards all other problems, they are as topical today as they were in those years. This circumstance testifies to the fundamental nature of the investigations carried out by the Jet Propulsion Research Institute.

Besides the study of liquid-propellant rocket engines, the RNII was also engaged in the investigation of air-breathing turbojet engines. The aircraft industry of those years was practically unfamiliar with this type engine which represented the future of jet-propelled aviation. It is to the institute's credit that the problems related to this engine figured prominently in its plans. The institute mainly engaged in developing the ramjet variant, which is still being used. The researchers concentrated on the process of combustion in ramjet engines and strove to improve their efficiency. The work was limited to subsonic engines. Theoretical studies conducted at the institute included investigations into the problems of turbojet engines, as well as into the principles of motorjet engines, now practically forgotten. Later the Institute also started work on designs of turbojet engines. All this convincingly shows that the RNII stood at the very source of the present-day problems in the theory and practice of aircraft engine construction.

As regards rocket technology proper, the RNII was actively engaged in the investigation into the possibilities of rockets which could be given today the name of ballistic missiles. The existence of such a trend was only too natural, though it was confined to the development of rockets for research purposes, mainly for meteorological studies. The reason lies in the fact that the thrust developed by engines in those years was not yet sufficient to lift heavy payloads, and the rockets therefore could hardly be used for combat purposes. Despite this fact, the development work on such rockets continued, and efforts were made to provide them with gyroscopic controls. On the whole, however, this work was of secondary importance.

By contrast with ballistic rockets, winged missiles enjoyed far greater attention. This was accounted for by fact that the weight of such missiles could be several times greater than the thrust of the rocket engine. Among such missiles were models 216, 212, 312, and 301 developed under Korolev's guidance. At that time Korolev was in charge of the team developing surface-to-surface (model 212) and air-to-air (model 301) winged missiles both of which were brought to the flight test stage. As distinct from earlier missiles, model 301 was not only equipped with gyroscopic autostabilizers, but also with radio guidance means, i.e., it already possessed all the basic elements of modern missiles of this class.

The development of the above missiles called for considerable effort to improve the automatic control system. The researchers developed gyroscopic instruments of the autopilot type suitable for operation under specific missile conditions. Understandably, they were made at the technological level of the 1930s and do not compare with modern gyroscopic automatic control instruments, if only because they were pneumatic, not electrical. In view of the difference between the technological levels, the automatic control of the 1930s rockets cannot be identified with those of modern ballistic rockets, yet it is highly significant that the former included all the key elements of automatic navigation systems. They comprised servo units, appropriate control programs, two gyroscopes and flight data automatic recorder system. As regards the theory of flight of winged missiles with automatic controls, the investigations in this field were far more advanced than the similar investigations in aviation. The theory of autopilot-controlled flight in aviation was practically at a standstill since the problems connected with such a flight were not very topical. Aircraft were mainly controlled by hand and their autopilots, unlike those of rockets, could be easily adjusted in flight. As regards rockets, their autostabilizers had to be adjusted on the ground without preliminary flight tests; therefore, great importance was attached to the theory of automatically controlled missile flight. In this field Korolev's group obtained essential results.

Alongside the development of flight machine designs and theoretical investigations directly linked with such development, the Jet Propulsion Research Institute carried out extensive research in supersonic wind tunnels. Its should be noted that the RNII was the only organization in the Soviet Union which had supersonic wind tunnels. They were employed for detailed investigation of jet nozzles with a view to optimizing their shape for subsequent use in jet engines. Significantly, already at that time, the researchers understood that it was desirable in some cases to heat the gas in the tunnel; therefore heated-air wind tunnels were developed. We shall not discuss here in detail the investigations carried out by the institute in the field of gas dynamics, since they were mainly aimed at obtaining the characteristics of projectiles, but not engines. These investigations were indeed very helpful in developing good rocket projectiles.

Besides design work, RNII members participated in research activity connected with rocket technology. For instance, the institute's scientific workers took part in the conference on stratosphere studies organized by the Academy of Sciences of the Soviet Union in 1934. This conference was mainly devoted to high-altitude balloons, but its agenda also included several reports on investigations of the stratosphere with the help of rockets. These reports were made by Korolev, Pobedonostsev and Tikhonravov. Another all-Union conference held a year later was devoted to the use of rocket vehicles in the investigation of the stratosphere. The reports were again made by Korolev, Glushko, Tikhonravov and Pobedonostsev. The RNII representatives thus participated in various scientific conferences at the highest level. The scientific workers of the institute also issued monographs which later became text books, and served as guides to rocketry for many enthusiasts who decided to follow their calling in life. Of such books I shall mention *Rocket Flight in Stratosphere* by Korolev, *Rockets, Their Design and Application* by Langemak and Glushko, and *Introduction to Astronautics* by Shternfeld. All of the authors were RNII workers.

In the period from 1936 until 1940, the scientific workers of the institute issued nine collections of articles on rocket technology in addition to RNII Transactions. This is convincing evidence that the RNII engaged not only in extensive design work and applied research but also carried out fundamental investigations. Its workers participated in scientific conferences sponsored by the U.S.S.R. Academy of Sciences, published books, issued collections of scientific articles summing up the results of their investigations, etc.

The brief survey of the RNII activity that I have made testifies to the broad range of research and scientific insight of its leaders: In point of fact, fifty years ago the institute concentrated on just those fields of rocketry which are in the focus of the scientists' attention today (rocket engines, autostabilizers, gas dynamics, rocket flight theory, etc).

Toward the end of World War II there appeared two tendencies that acted in opposite directions. On the one hand, the institute began to expand owing to the influx of fresh forces; on the other hand, it started dividing into independent research organizations.

Groups of scientists who had begun to study the problems of jet propulsion in other centers joined the RNII. For instance, such scientists as Sedov, Petrov, Abramovich, Pilugin and others transferred to the RNII from the Central Aerohydrodynamics Institute (TsAGI). Another body that joined the institute was Bolkhovitinov's aircraft design bureau, which had developed the rocket interceptor plane BI-1. A number of combustion theory specialists, including Gukhman and Knorre, came from the Leningrad Central Boiler-Turbine Institute. In 1946 the institute was joined by Keldysh. Hence, the RNII drew all those who were concerned with rocket technology. However, in view of the rapid expansion and specialization of its activities, the institute, which was the sole and universal organization in the highly sophisticated and extremely important field of jet propulsion started breaking up into specialized institutions - one for projectiles, another for liquid-propellant engines, a third one for control problems, a fourth one for ground facilities, etc. At different times the RNII gave birth to two independent design bureaus - one headed by Isayev which later played an important role in the development of Soviet space technology, and the other one headed by Lulka that was concerned with aviation problems. It is appropriate here to mention that the first Soviet turbojet was developed in Lulka's bureau. Another independent design organization born within the institute was Bondaryuk's bureau concerned with the ramjet design.

The history of the Jet Propulsion Research Institute in its initial organizational framework lasted but little more than a decade. The institute was set up in 1933, and ceased to exist in its original form in 1945-1946. Its scope began to narrow and it gradually lost its significance as a universal research center representing such diverse fields as artillery, rocket design, liquid-propellant rocket engines, automatic control systems, gas dynamics. This center had played a unique role in the development of Soviet aerospace investigations. Its leaders, such as Kleimenov, Langemak, and Korolev clearly understood that all the problems tackled by the Jet Propulsion Research Institute were important, and that it was impossible to develop, for instance, rocket engines or automatic controls without paying due attention to other fields. As soon as one or another trend gained sufficient strength and was able to sail on its own, no one attempted to prevent its separation from the institute - on the contrary, such an event was always welcome.

Modern space industry is an extremely complex field linked with practically all branches of science and technology. Looking back now at the history of the Jet Propulsion Research Institute, one can plainly see that all the trends of modern rocketry have their roots in that glorious decade. The history of the RNII, conceived and set up as the center of Soviet rocket studies in the early 1930s, came to an end with the termination of World War II. It was indeed a short but glorious history.

AAS 91-284

Chapter 4

# THE BRITISH INTERPLANETARY SOCIETY: THE FIRST FIFTY YEARS (1933-1983)[*]

G. V. E. Thompson[†] and L. R. Shepherd[‡]

**BACKGROUND**

Until the mid-1920s serious scientific speculation on the future of spaceflight had been the province of individual scholars such as Tsiolkovskii, Goddard, Oberth, and Esnault-Pelterie, but then a new phase in man's extraterrestrial aspirations was opened by the formation of groups devoted to the conquest of space.[1] The first of these appears to have been the Society for the Study of Interplanetary Communication (OIMS), which was founded in Moscow in June 1924 under the chairmanship of G. M. Kramarov and included among its registered members such well-known figures as Tsander, Perelman, and Tsiolkovskii himself.[2] The OIMS was short-lived, passing from sight after only one year.

Three years later, in Germany, the Verein für Raumschiffahrt (VfR) was formed on the 5 July 1927 at a meeting in the Wirthaus zum goldner Zepter (Golden Sceptre tavern) in the town of Breslau.[3] This society flourished; its membership, which included such celebrated pioneers as Oberth, Hohmann, Esnault-Pelterie, and von Pirquet, grew rapidly to reach about 1000 within two years. It published a journal, *Die Rakete*, which printed a wide range of articles of both popular appeal and scientific interest. Oberth succeeded Winkler as its president and then, early in 1930, its headquarters moved to Berlin, where it embarked upon a series of experiments on liquid-propellant rockets at its famous Raketenflugplatz at Reinickendorf. However, after 1929 it was forced to discontinue publication of *Die Rakete*, the membership quickly melted away, and by the end of 1933 the VfR had broken up. The decision to concentrate its resources on rocket experimentation contributed to the demise of *Die Rakete*, and the consequent lack of communication with its members was fatal to the society.

---

[*] Presented at the Seventeenth History Symposium of the International Academy of Astronautics, Budapest, Hungary, 1983.

[†] Immediate Past-President, British Interplanetary Society, Vauxhall, London, United Kingdom..

[‡] Former President; British Interplanetary Society, Former President IAF.

During the short era of the VfR other rocket and space societies emerged in Austria, France, and the U.S.A., the most notable of which was the American Interplanetary Society (later renamed the American Rocket Society), which came into being on 4 April 1930. Then on Friday, 13 October 1933, at a meeting in Liverpool, a British society was formed which was to share with the American Interplanetary Society the distinction of surviving into the Space Age.[4-6]

## FORMATION

On 8 September 1933 Philip E. Cleator published in the *Liverpool Echo* newspaper an appeal for members to form a British interplanetary society. The result was disappointing, the appeal eliciting only one reply. However, one of Britain's national newspapers, *The Daily Express*, sent a special correspondent, Moore Raymond, to talk to Cleator. Raymond was an immediate and enthusiastic convert to the idea of the proposed society and secured a part of the front page of the *Express* on the following morning to publicize the proposal. This time the response was considerable and within a week Cleator was able to hold, in his Wallasey, Cheshire, home, a small meeting of enthusiasts who lived nearby. At that preliminary gathering the decision was taken to form the British Interplanetary Society, and an inaugural meeting was scheduled to be held in Liverpool on Friday, 13 October.[7,8]

The meeting duly took place on the day specified at offices in Dale Street, Liverpool. The British Interplanetary Society (BIS) was founded, with Philip Cleator as its first President. He urged that a printed journal would be essential to the new society. This was agreed and he was asked to edit the publication. From the outset the *Journal of the British Interplanetary Society* (or *JBIS*, for short) was to be a glossy illustrated publication, at first no more than a six-page pamphlet, but by the time it had reached its last pre-war issue, Volume 5, No. 2, in July 1939, the periodical had increased to a respectable size with 28 pages, about the same as *Astronautics: Journal of the American Rocket Society*, which had an identical format (140 mm x 220 mm).

*JBIS* was then published at intervals of about six months and it was always recognized that such a low frequency of publication was not conductive to growth. Already in 1935 G. E. Pendray had suggested that the journals of the American Rocket Society (of which he was President), the British Interplanetary Society, and the Cleveland Rocket Society should be merged to form a single monthly publication. The estimated cost, however, proved to be too high for the BIS, so the proposal was shelved.

The rate of growth of the BIS was quite modest by comparison with the VfR. At the end of the first three months the membership numbered only 15, and it had not quite reached 200 when activities had to be suspended at the outbreak of World War II in September 1939. At an early stage the Society attracted to its ranks the inevitable cadre of distinguished pioneers, notably Esnault-Pelterie, Pendray, von Pirquet, Perelman, Rynin, and Willy Ley, a pioneer of the VfR, who had become a Fellow of the BIS soon after its foundation.

With the growth of the BIS there was a shift in the center of gravity of the Society away from the Liverpool region towards London. This was inevitable, since a quarter of the British population lived either in the capital or within easy commuting distance. The membership in this area soon outnumbered that in Liverpool and its environs and it became increasingly apparent that there was an overwhelming case for moving the headquarters to London,[9,10] after dissension in the Liverpool-based BIS Council had led to Cleator tendering his resignation from that body and relinquishing the editorship of the *Journal*, in April 1936.

Later in the year, at a meeting of London members at the Piccadilly offices of the Vice-President, A. M. Low, on 27 October 1936, the London Branch of the BIS was established. By this time, Society affairs were in crisis with the organization at the Liverpool headquarters rapidly falling apart. The climax was reached on 6 December at a General Meeting of the Society, the last to be held in Liverpool. The London Branch was invited to take over and the last act of the Liverpool group was the publication of the February 1937 issue of the *Journal*.

The parallel with events in the VfR, in 1929/1930, when that body transferred from Breslau to Berlin, can scarcely escape attention. The outcome for the BIS was fortunately happier.

The Liverpool period of the BIS was, perhaps, the romantic era of the Society. For the three years that it lasted it was dominated by Phil Cleator. He was President, Editor, and author. His book[7], *Rockets through Space*, was one of the best popular publications on space travel that emerged in the interwar years, and, no doubt, stirred the imagination and interest in astronautics of many a newcomer to the subject. That the BIS survived those first delicate years from its birth was due largely to the fact that Cleator spared no effort to publicize it, nor in his drive to recruit new members. In the end, at Liverpool, his domination was resented and this led to his resignation from the presidency.

## MOVE TO LONDON

With the move to London, A. M. Low (Vice-President) stepped up to take over as the second President of the BIS. The Liverpool connection was not entirely severed for, with good grace, Cleator and L. J. Johnson (hitherto Honorary Secretary) accepted the positions of joint Vice-Presidents.

The transfer to the capital naturally produced some interruptions in Society affairs and in this respect publication of the *Journal* suffered most, with the December 1937 issue appearing well into 1938 and there being no issue bearing a 1938 dateline. In fact, only three issues of the *Journal* were to emerge from London in the two-and-a-half years that remained before war interrupted the functioning of the Society. By way of compensation, these London publications were of much greater substance technically than their Liverpool predecessors.

One of the first actions in London, under a revised constitution, was to set up a Technical Committee with a Research Director. The BIS was seeking an appropriate role in the development of space travel and with this Committee it soon found one. Many members still thought enviously of the VfR and its Raketenflugplatz and of pursuing an experimental program to develop the hardware of rocketry. But this was not realistic. An objective appraisal of the situation was bound to show that the development of rocket engines and vehicles was far beyond the capabilities of any spaceflight society. In reality, the experimental activities of the VfR and of its American successors contributed nothing of significance to rocket technology. They were little more than games, the real business calling for resources that could only be provided by government agencies or large industrial organizations with well-funded research institutes. It is important in stressing this fact that one should not denigrate the work of the VfR, however, because the activities of the Raketenflugplatz brought the rocket to the attention of people who were to have at their disposal the resources that were needed to pursue the matter in earnest. Moreover, even if that activity did not lead to any notable achievement in launching rockets, it was successful in launching the careers of future rocket engineers of the caliber of von Braun. Nevertheless, by the late 1930s the game had been played and experimentation in rocketry had moved on to a higher plane and was not an option open to the BIS.

But there was an option open to the Society and this lay in the conduct of technical assessments and conceptual design studies. Such activities were and are an essential feature of the development of all technology. The BIS Technical Committee chose to follow this course and embarked upon its now famous lunar spaceship study. The object of the exercise was to produce a conceptual design of a vehicle of modest initial mass (1000 tons) which would be capable of placing a 1 ton payload including two astronauts on the surface of the Moon together with sufficient propulsion capacity to return that payload into the hyperbolic orbit intercepting the Earth atmosphere. Aerodynamic braking and parachute descent would take care of the safe descent of the astronauts to the surface of their home world.

In order to achieve the necessary payload ratio, the BIS study group adopted a principle of continuously discardable structure in which the propulsive stages of the vehicle were subdivided into about 2500 small units, each made up of a rocket engine and propellant. Sufficient of these would be fired at any time to provide the required thrust and each would be jettisoned immediately. Since no restart would be required in most of these units the advantage of using solid-propellant motors was recognized. The BIS group referred to this form of construction as their "cellular" design. In fact it was the principle which Goddard had conceived 20 years earlier in his famous paper "A Method of Reaching Extreme Altitudes".

· In this particular aspect of the lunar vehicle concept the BIS designers (and Goddard, for that matter) turned out to be quite wrong. They did not anticipate the enormous advances that would be made 15 to 20 years later in structural design and in the development of very large rocket engines of high thrust/mass ratio. However, they were designing as close as possible to the state of the art at that time, when even their relatively small 100,000 HP (75,000 kw) solid-propellant motors represented a significant extrapolation in the technology. At least it can be said that the BIS group recognized the value of high-performance solid-propellant motors in spaceflight, when continuous single-burn programs were to be involved.

If the main propulsion features of the BIS Lunar Spaceship turned out to be far from the mark, then in many other respects the opposite was true. The study group, for example, appreciated the requirement of inertial navigation devices and, indeed, proposed some experimentation on these. However, it was in the matter of the lunar landing and subsequent return from the lunar surface that they showed their clear understanding of the problem and its solution. In particular, they sketched out the main requirements of the lunar landing vehicle. The intervention of the war prevented the filling-in of detail, but in the immediate post-war years R. A. Smith updated the lunar landing vehicle concept and his design (August 1947) anticipated the Apollo Lunar Excursion Module in every essential feature, including the use of the base section as a launching platform.

## WORLD WAR II

On 3 September 1939 Britain entered World War II, and the BIS Council, recognized the impossibility of continuing to function effectively in the metropolitan area, decided to suspend operation of the Society for the duration of hostilities. It was the end of the pre-war era of the BIS that had extended over six years. In just that length of time the VfR had been born, had risen swiftly to glory, and then faded suddenly from sight. The BIS had a much more modest entry into the world of astronautics and its progress to maturity had been slower and far less spectacular. But by 1939 it had been set on a sound course that was to take it surely in the right direction.

During the war years some members of the BIS maintained communication and also established contact with other fledgling astronautical societies in Britain.[11] Towards the end of the period of hostilities two of these had come together to form a group called the Combined British Astronautical Society (CBAS) whose leaders then expressed an interest in merging with the older BIS.

## REAWAKENING AND GROWTH

On 13 June 1945 the BIS re-emerged from its six-year hibernation with an informal meeting convened by A. M. Low at the Royal Automobile Club in London. Only three of the old Council, namely, Low, Cleator, and R. A. Smith, were present, so an emergency committee of these three and seven other members present was appointed to act for the Council. The main decision taken by the emergency committee was that the BIS should be incorporated and its constitution revised to conform with the legal requirements for this. L. J. Carter undertook to draft a revised constitution. The implementation of the decision was overtaken by events, for on 25 September representatives of the BIS and CBAS met and agreed to merge to form an enlarged British Interplanetary Society, which should then be incorporated. Carter continued with his appointed task, preparing the Memorandum and Articles of Association. These were completed and then approved at a further meeting of representatives of the merging groups and on 31 December 1945 the British Interplanetary Society received the certificate of registration as a company limited by guarantee and having no share capital.

L. J. Carter was appointed Honorary Secretary of the Society, an office which was eventually discontinued to give way to the post of Executive Secretary. To the enormous benefit of the Society, this incumbancy has remained in his capable hands since that time. The new Council consisted of twelve members, including a Chairman, and eventually a Vice-Chairman. There were no officers other than a Technical Director (a continuation of the pre-war post of Research Director). The office of President was allowed to lapse and was not re-introduced until 1960 (Table 1). In the interim, the Chairman of the Council acted in the presidential capacity and, retrospectively, the Council Chairmen between 1946 and 1959 are now rightly regarded as past BIS Presidents.

Table 1
PRESIDENTS OF THE BRITISH INTERPLANETARY SOCIETY

| PRESIDENT | PERIOD OF OFFICE |
|---|---|
| P. E. Cleator | 1933-1936 |
| A. M. Low | 1937-1939 |
| E. Burgess[a] | 1946 |
| A. C. Clarke[a] | 1946-1947 |
| A. V. Cleaver[a] | 1948-1950 |
| A. C. Clarke[a] | 1951-1953 |
| L. R. Shepherd[a] | 1954-1956 |
| R. A. Smith[a] | 1956-1957 |
| L. R. Shepherd[a] | 1957-1960 |
| W. R. Maxwell | 1960-1963 |
| M. N. Golovine | 1963-1965 |
| L. R. Shepherd | 1965-1967 |
| W. R. Maxwell | 1967-1970 |
| G. V. Groves | 1970-1973 |
| K. W. Gatland | 1973-1976 |
| G. V. Groves | 1977-1979 |
| G. V. E. Thompson | 1979-1982 |
| A. T. Lawton | 1982- |

[a] In the period 1946-1960 the most senior officer of the Society, although undertaking the duties of President, was known as Chairman of the Council.

For a few months after its incorporation the Society was limited to publishing a duplicated *Bulletin*, but as membership grew it became possible to have this publication printed. Publication of a printed *Journal* was also resumed and an *Annual Report* was supplied to members. For a year or two all three periodicals appeared, until it was decided to concentrate the Society's efforts on the senior publication. So the *Bulletin* and *Annual Report* were dropped, after a useful but rather short existence.

The journals now being issued were much more professional publications than the pre-war issues, both as regards content and production, and their standard has continued to improve. Mathematical papers on subjects such as interplanetary flight and rocket combustion began to appear. An important source of contributions to the *Journal* was the series of monthly lectures organized by the Society, held first at the St. Martin School of Art and later at Caxton Hall, Westminster, but many others were specially written for publication. Only a few papers can be mentioned here: "The Atomic Rocket", by L. R. Shepherd and A. V. Cleaver, "Interplanetary Man" by Olaf Stapledon, "Orbital Bases" by H. E. Ross (dealing with the establishment and design of spacestations), "Lunar Spacesuit" by H. E. Ross, "Perturbations of a Satellite Orbit" by Lyman Spitzer, and "The Man-Carrying Rocket" by R. A. Smith. The latter was a proposal for converting a captured V2 rocket to carry a man, to enable the first manned sub-orbital spaceflights to be made. In addition to being presented to the Society as a paper these proposals were put formally before the Air Ministry, but no action resulted. More than a decade was to elapse before Shepard and Grissom made such flights with a Redstone rocket in the U.S.A. Of course their flights were less hazardous than that envisaged by Smith but volunteers were available to make the attempt and the dangers are of much the same magnitude as in crossing the ocean single-handed in a rowing boat or a barrel--almost commonplace voyages today!

A series of composite papers presented by K. W. Gatland, A. M. Kunesch, and A. E. Dixon in 1950-51 "Initial Objectives in Astronautics", "Orbital Rockets", and "Minimum Satellite Vehicles" considered useful astronautical projects,[12] short of manned spaceflight, which could be accomplished with the then existing rocket propellants and engineering practice. The third paper had four schemes for vehicles which should be able to put into orbit around the Earth minimum satellite payloads capable of providing useful scientific information. A connection has been traced from this work to the first U.S. artificial satellite.[13]

In the immediate post-war years the pursuit of technical assessment was continued, with a Technical Committee under a Technical Director. Study Groups dealing with various aspects of spaceflight were set up and these met from time to time. However, there was no single project, such as the pre-war Lunar Spaceship, on which attention could be focused and the work of the groups gradually evolved into efforts of individual enthusiasts or of partnerships of two or three people who pursued their favored topics without direction. Eventually the need for a technical committee and director disappeared because undirected ventures were proving to be highly fruitful in producing original *Journal* papers.

## FOUNDATION OF THE IAF[14,15]

While the BIS was consolidating its position in the UK and also attracting many overseas members, a number of astronautics and rocket societies were being established in other countries. By 1950 the BIS was in touch with bodies in Argentina, Denmark, Germany, Canada, and the U.S.A., including the pre-war American Rocket Society. In the Federal German Republic, three autonomous groups made up the Gesellschaft für Weltraumforschung (GfW). Close contact existed between the British society and the Stuttgart GfW, when in June 1949 the Board of Directors of the Society passed a resolution calling for an international meeting of all societies concerned with rockets, interplanetary travel, and space research, in order to foster friendly relations and the exchange of information and to explore the possibilities of forming an international astronautical association. The GfW communicated this resolution to other societies and asked the BIS if it would be prepared to hold the proposed meeting in London. The resolution was well received by the BIS Council and a reply indicating its broad agreement was sent to the West German society.

Correspondence developed between the GfW, the BIS, and the Groupement Astronautique Français on the proposed international association and the meeting of the various national societies that would form it. The BIS considered that two years would be required to set up a full-scale London conference, but M. Alexandre Ananoff, the President of the French group, indicated that he would be able to organize a preliminary meeting in Paris. This Paris meeting was held on 30 September - 2 October 1950. It was limited to a large public meeting on the first day and small business meetings of the societies' representatives on the other two days. The public meeting was impressive, being held in the Richlieu Grand Amphitheatre of the Sorbonne with about 1000 people in attendance. H. Mineur, Director of the Institut d'Astrophysique, presided and many distinguished public figures were present. This type of opening meeting was to form the pattern for subsequent Congresses.

The business meetings were held at the French Aero Club and were confined to the official representatives of eight participating societies and four independent observers. They were (in addition to Ananoff): Tabanera (Sociedad Argentina Interplanetaria); Cap, Rückert, Schmiedl (Oesterreichische GfW); Hansen (Dansk Interplanetarisk Selskab); Brugel, H. H. Koelle, Loeser (GfW, Stuttgart); Jungklass. Oesterwinter (GfW, Hamburg); Mur (Agrupación Astronautica Española); Hjerstrand (Swedish Society in process of formation); Burgess, Clarke, Cleaver, Humphries, Shepherd (British Interplanetary Society). The independent observers were Frau Dr. Bredt, Engels, Nebel, and Sänger, providing a nostalgic link with the old VfR.

These plenary meetings did not establish the international association, but at the end six points were resolved unanimously:

1. That such a body should be created.

2. That it should be inaugurated at a Congress to be held in London in 1951;

3. That in the interim, representatives of the individual societies should correspond, to exchange views and proposals relating to the proposed astronautical federation.

4. That the British Interplanetary Society, which would organize the 1951 London meeting, be charged with the task of co-ordinating the proposals submitted.

5. That the functions and constitution of the proposed international body should be worked out well in advance of the London meeting and should form the basis for the discussions at the plenary sessions, leading to an agreement to inaugurate the international organization.

6. Pending the setting-up of the international body, Dr. Eugen Sänger would preside over a provisional committee consisting of the leaders of the delegations present at the Paris meeting.

In the intervening year there was a considerable exchange of correspondence, most of it between the Stuttgart GfW and the BIS and indeed it owed much to the efforts of Guenter Loeser of the BIS that the British society was able to circulate a draft Constitution of the IAF in May 1951, four months ahead of the Conference.

The Second International Astronautical Congress was held in London during the week commencing 3 September 1951 with the BIS acting as host. It should be noted that the Congresses are numbered from the Paris meeting and that consequently the Congresses predate the International Astronautical Federation (IAF) by one year, a fact that is not always appreciated. The venue was Caxton Hall, Westminster, which by this time was being used extensively by the BIS for its various meetings. Some 40 overseas visitors were present at the Congress, which was also well attended by members and others from Britain. At the plenary sessions in addition to the Societies that had been represented in Paris there were delegates from astronautical bodies in the Netherlands, Italy, Switzerland, and the U.S.A. Among those present was Hermann Oberth.

The London Congress was the prototype of the annual meetings that were to follow in having a public meeting, plenary session of delegates, and a technical symposium. The symposium held at London had as its theme "Earth-Satellite Vehicles," a down-to-Earth topic which properly emphasized the first step in spaceflight. At that time there was no need to hold parallel sessions and there was no international program committee, the symposium being co-ordinated by the host society (in the case of the London Congress the BIS Technical Advisory Committee[*] was responsible).

A. C. Clarke, on taking over as Chairman of the BIS Council from A. V. Cleaver, welcomed those attending the Congress and Eugen Sänger presided over the plenary meetings of the delegates. The International Astronautical Federation was duly inaugurated at the end of these sessions on 4 September 1951. The constitution drafted by the BIS on the basis of its correspondence with its contemporary founder-societies was not immediately adopted, however, because of disagreement on a single issue (the question of voting). It was left to the Third International Congress to agree to the principle of having a single voting society in each country represented in the Federation. This is still a controversial issue.

The BIS is naturally proud of the major role that it played in the foundation of the IAF. It is a matter of interest that at the 10th International Astronautical Congress (also organized by the BIS in London) in 1959 the International Academy of Astronautics (IAA) and International Institute of Space Law (IISL) came into being.

## PERMANENT HEADQUARTERS

Another important development occurred in 1952. On 12 May 1952 the Society moved into an office (a modest L-shaped room) at 12 Bessborough Gar-

---

[*] Technical Director (L. R. Shepherd), Deputy Technical Director (G. V. E. Thompson), and Council Chairman (A .V. Cleaver).

dens, London SW1 as tenants of the Queen and her government--or more precisely, of the Crown Agents. This was a normal commercial transaction--the BIS is an independent Society neither receiving a government subsidy nor subject to any government control or direction (other than the law of the land, which applies to all citizens and corporate bodies). The Society does, however, have excellent relations with government scientists and establishments working in relevant fields. This is often a two-way relationship, with each party doing what it can to assist the work of the other.

About this time, a paper was read on "Interstellar Travel", demonstrating that the Society's interest in space had no limitations. But more pressing problems were not neglected. The design and testing of rocket motors received considerable attention, as did the mathematics of transfer between different orbits, food and atmosphere control in space vehicles, landing on airless planets, a lunar base, and a host of other topics. Demonstrations of the properties of rocket propellants were given and visits to various research establishments were organized for members.

Besides these "internal" activities, leading members of the Society worked hard to disseminate information about astronautics to the general public by means of books, magazine and newspaper articles, occasional radio programs, but above all to scientific or social organizations, schools, factories, trade unions, youth groups, etc. Common fallacies and misconceptions about interplanetary travel were gradually dispelled, interest in the subject was awakened, and as a result the Society's membership rose rapidly in the early 1950s.

Rockets were now being used extensively to explore the upper atmosphere. In August 1953 an important conference on this subject was organized by the Royal Society at Queens College, Oxford. About half the BIS Council attended and participated in the discussion. The most significant of the papers presented was that of Singer, who developed the ideas of Gatland and co-workers (mentioned above) and proposed the MOUSE satellite vehicle (Minimum Orbital Unmanned Satellite Earth) concept and identified many of the applications for which the 100 lb satellite would be suitable for studying solar radiation, cosmic rays, weather, etc.

Although the Society's eyes were directed firmly towards the future, the past was not neglected. As World War II receded into history, secrecy was relaxed and it became possible for the *Journal* to publish valuable accounts by German engineers, such as Riedel and von Braun, describing early rocket experiments and the development of the V2 (or A4) and other rocket weapons. Data sheets were printed on existing rocket vehicles and engines.

It is impossible to mention here all the topics covered in the Society's *Journal* at that time--suffice to say that they dealt with all aspects of spaceflight, often in considerable detail. Some have since become matters of fact; others (such as extraterrestrial mining or farming) have still to be put into practice. Two series of mathematical papers--those of Lawden on the correction and perturbation of interplanetary orbits, and those by King-Hele on the descent of Earth satellites through the upper atmosphere--were providing the theory which would be needed in all the forthcoming work with artificial satellites.

In July 1955 the U.S. Government announced an artificial satellite program which would be linked with the International Geophysical Year (IGY), due to commence in July 1957. The BIS began to organize "Moonwatch" and radio tracking teams to observe the satellites (it should not be forgotten that this was before the establishment of the massive governmental and international tracking stations--the giant radio telescope at Jodrell Bank was in construction but had run into engineering and financial difficulties). At the Copenhagen Congress of the IAF (August 1955) the U.S.S.R. also announced interest in artificial satellites, but little information about their plans was given.

This led to an increase in the public awareness of astronautics and so the BIS Council judged the time right for the issue of a popular space magazine. In October 1956, *Spaceflight* commenced publication, initially under the editorship of Patrick Moore, the well-known amateur astronomer, author, and TV personality. A selection of twenty-four of the best papers from the Society's *Journal* also appeared as a book edited by Len Carter[16] and was an immediate success. Len Carter had edited the *Journal* since March 1947 as well as being the Society's Secretary, but in view of the great increase in office he now had to relinquish the editorship, G. V. E. Thompson replacing him.

After the war the membership of the BIS grew steadily to reach just over 900 in the first five years after its incorporation. Then in the space of two years it more than doubled, standing at 2010 on 30 September 1952. To some extent this spectacular increase reflected the considerable attention given by the national press to the Paris and London Congresses.

The BIS now matched the ARS in the size of its membership and shared the honor of being one of the two largest astronautical societies in the world. The progress in membership continued, passing the 3000 mark in 1957 and reaching a peak of 3300 at the end of 1959. There was a slight drop in membership during the mid-sixties but by the mid-seventies it had stabilized again at just over 3000. This level sometimes appears to have been fixed by some natural law!

Other societies and educational establishments were now finding that certain aspects of astronautics were having an impact in their own fields, so it was opportune to hold joint symposia with the BIS on appropriate topics. The first of these was the Symposium on High Altitude and Satellite Rockets,[18] organized jointly by the BIS, the College of Aeronautics, Cranfield, and the Royal Aeronautical Society and held at the College in July 1957. Details of the U.S. satellite program were given and a Russian delegation attended.

## DAWN OF THE SPACE AGE

On 4 October 1957 as delegates (including a strong BIS contingent) were travelling to Barcelona for the eighth IAF Congress, there came the news of the launching of the first Russian Sputnik. The Space Age had begun! Astronautics immediately captured the imagination of press and public, for what had previously seemed fantasy had become fact. Activity increased in all aspects of BIS work.

In 1958 the Society was twenty-five years old. This special anniversary was celebrated at a banquet at the Waldorf Hotel, London, held in conjunction with the International Symposium on Space Medicine and Biology organized by the BIS and held in the Great Hall of the British Medical Association.[19,20] The voyage of the dog Laika in the second Russian satellite, *Sputnik 2*, on 3 November 1957 had indicated that manned spaceflight would not be long in coming and so this symposium was well attended. Three of the papers presented were from the RAF Institute of Aviation Medicine and a strong team from the U.S. Office of Naval Research discussed American work. The early unmanned satellites had given valuable results for this field, one interesting discovery being the van Allen radiation belts around the Earth.

The following year (1959) was an exceptionally heavy one for the BIS. In addition to its ordinary activities of publishing and local meetings, the BIS was again host-Society to the IAF and organized the Tenth International Astronautical Congress at Church House, Westminster. The Society at last received a good measure of Government support, with the Congress being opened by the Minister of Supply and the delegates being invited to a reception at Lancaster House. Eighty-one papers were read at the Congress. In addition, there was a Space Law Colloquium held in parallel at Lincoln's Inn. The whole series of meetings was preceded by a Commonwealth Space Flight Symposium, which surveyed work then going on in the British Commonwealth.[21] Proposals were made at the Symposium for the establishment of a joint Commonwealth Space Agency, but these came to naught.

## BIS MEMORANDA TO UK GOVERNMENTS

After the launching of the first artificial satellite the BIS increased its activity in the political arena, being concerned to see a strong industry devoted to space technology established in the U.K. Leaving aside any question of national pride, this seemed to be clearly in the interest of the Society in securing its future professional standing. At this time, although lagging behind the U.S.A. and U.S.S.R., Britain was seriously in the business of rocket propulsion, having at an advanced stage of development a large liquid-propellant rocket (Blue Streak) and a smaller one (Black Knight). These had been intended for a military application, which was abandoned, and the opportunity existed of adapting them as spacecraft launching vehicles. The BIS was anxious to support this.

In January and February 1960 two special meetings were convened by the Society with the object of drawing up a memorandum concerned with the need for a U.K. space program and intended for submission to the Government. The participants in these meetings comprised BIS Council members, senior members of the aeronautical and space-related industry, and political figures. Most of those present were members of the BIS. The document produced made a number of recommendations under the heading "A Space Programme for the UK", the main features of which were:

1. A specific project to develop Blue Streak and Black Knight as a satellite launching vehicle.

2. A later project to develop a more powerful launch vehicle combining Blue Streak with higher-performance second and third stages.

3. A detailed feasibility study of communications satellites and an examination of their significance.

4. A program of research on winged reentry vehicles.

5. A modest program on space medicine.

6. A long-term program on advanced propulsion systems (nuclear, electrical, etc).

The above Memorandum was submitted to the Prime Minister (Harold Macmillan) on 2 March 1960, and was referred to the Science Minister (Lord Hailsham), who received a deputation from the BIS (A. V. Cleaver, P. Masefield, L. R. Shepherd) on 10 May. The outcome of the meeting was that the BIS was invited to submit a more detailed document to the Prime Minister and the Minister for Science stressing the benefits to the country of a space program and its impact upon technology in general. This the Society duly did.

Subsequently, BIS representatives met an all-party group of MPs at the House of Commons to assist them in setting up a Parliamentary Committee on Space.

The BIS proposal to develop a launch vehicle based upon a combination of Blue Streak and Black Knight was never implemented, though it was the subject of a feasibility study carried out in the Royal Aircraft Establishment at Farnborough. However, the U.K. Government submitted a proposal to other European countries which led to the setting-up of ELDO, with the principal object of developing a three-stage launching vehicle (Europa) using Blue Streak as the first stage. This project, unfortunately, was abandoned some years later after the U.K. reneged on the agreement with its ELDO partners during the term of the Wilson Government. It may be realized, however, that a smaller launching vehicle, Black Arrow (based on Black Knight) was developed, at a very low cost, and used to launch a small satellite. This too was abandoned after one successful mission and subsequently the British aerospace industry concentrated on spacecraft and played no part in developing launchers.

In 1965 the BIS submitted another detailed memorandum to the Wilson Government on the theme of European cooperation in space, the main points being:

1. The need to set up a single Western European space authority.

2. The need for Western Europe to develop more powerful launching vehicles, including a reusable type.

3. The undertaking of a broad program of space research covering both the Earth's environment and deep space studies.

4. The pursuit of a manned spaceflight program.

This memorandum met with little response, the U.K. Government of the day being unfavorably disposed towards space technology and to some extent hostile to European collaboration in the field. The Government was responsible for the withdrawal of British support from ELDO (which had been set up on Britain's initiative) and for moving the country down from the fairly forward position that it had hitherto occupied in space technology to a much lower status.

Seven years later, in February 1972, the BIS sent a further recommendation to the Heath Government, again urging the establishment of a European Space Authority. This time the Society was in powerful company, for the U.K.'s Science Research Council and Aeronautical Research Council were of the same opinion on this issue and were also pressing for a European counterpart of NASA. The British Minister for Aerospace and Shipping (Heseltine) put this proposition to a European Space Conference in November 1972 and on 1 January 1974 the European Space Agency was founded.

## SYMPOSIA

After the Tenth International Astronautical Congress it was decided that the holding of technical symposia should be a regular feature of BIS work; and that practice has continued to this day. On occasion they are held in conjunction with other organizations. The time seemed opportune for a closer grouping of the European astronautical societies and it was decided to hold periodic European Symposia on Space Technology. The BIS hosted the first of such meetings, held at Federation of British Industries House in June 1961. Similar symposia have been held ever since then, with the venue rotating from country to country.

The titles of the subjects chosen for the Society's own one- or two-day symposia during the early 'sixties showed how widely these ranged over most aspects of astronautics--the topics ranged over Communication Satellites, Liquid Hydrogen as a Rocket Propellant, Rocket and Satellite Instrumentation, Space Navigation, Materials in Space Technology, Navigation and the Early Exploration of the Moon, Astronautics in the School Curriculum, Generation of Power in Space, Meteorology from Space, Advanced Propulsion, Aerospace Vehicles, Ground Support Equipment, the ELDO Launching Vehicle, and the Engineering of Scientific Satellites. The papers presented and the ensuing discussion appeared in the Society's *Journal*-- sometimes they even reached a wider readership by being also published in books.[20-27]

As part of its work in the field of education, the BIS organized a course on Rocket Motor Technology, intended for teachers. A *Teacher's Handbook of Astronautics* plus two supporting books and materials were published and proved popular.[28]

One of the Society's duties has been, and still is, to signal its recognition of the achievements of distinguished workers in astronautics by making appropriate awards. A grade of Honorary Fellowship was used to honor the work of the early pioneers. When spaceflight really came about, it was deemed appropriate to extend the range of awards to gold and bronze medals for individuals and silver plaques for group achievements. The first gold medals were awarded in 1961 to Yuri Gagarin (for the first manned flight in space) and to Wernher von Braun (for the development of the Pershing, Redstone, Jupiter, and Saturn rockets at NASA's George C. Marshall Space Flight Center). The third was awarded in 1964 to Valentina Nikolaeva (*née* Tereshkova), the first woman to make an orbital flight. Both the Russian cosmonauts were able to come to London at the invitation of the BIS to receive their medals and many members attended the ceremonies. Subsequently, Edwin Aldrin, Neil Armstrong, and Michael Collins each were to receive a gold medal for the Society in recognition of the triumph of *Apollo 11*, while NASA was presented with a silver replica of the lunar landing vehicle in recognition of the group effort. Other silver awards have recognized both American and Russian successes in space, but are too numerous to mention.

After much discussion in Council and with the Society's solicitors, a new Constitution (or rather Memorandum and Articles of Association) was approved in 1965. This included abandoning powers to carry out mining and other operations on the Moon, etc. About that time the decision was taken to suspend publication of the *Journal* temporarily in order to place *Spaceflight* on a monthly basis from January 1966. K. W. Gatland had replaced Patrick Moore as Editor of *Spaceflight* in 1959 and was to continue in that office until May 1981. *Spaceflight* has continued to appear monthly ever since 1966, except for occasional double numbers made necessary by staffing problems at holiday times.

In 1966 the BIS received a Spacemobile on loan from NASA for several months. This was a very large van, fully equipped with a wide variety of models, films, and other visual aid equipment. Lecturers from the Society toured Britain with the vehicle to carry out experiments and demonstrations at hundreds of schools, colleges, teachers' training centers, factories, and service establishments.

By 1968 the finances of the Society had improved and it became possible to recommence publishing the *Journal*, at first quarterly and later bi-monthly. Professor G. V. Groves was appointed Editor. The new Constitution now enabled the Society to be registered as a Charity, which gave it a number of tax advantages. An equipment fund was launched, which allowed composing and ancillary equipment to be bought, with the result that great savings were made in typesetting *Spaceflight* (from 1972) and the *Journal* (from 1974). They were typed out and paged in the Society's Headquarters, although still printed externally. The *Journal*, meanwhile, became a monthly again in 1970. Its page size had been reduced when it reappeared, but in 1978 it was restored to its old format (A4). These changes provided a greatly improved service to members but meant that the work at the office was multiplied by an even greater factor. More working space was urgently required especially since the threat of demolition of the Society's offices was looming to make way for redevelopment of the surrounding area, mostly Crown property.

**Figure 1**  Old BIS Headquarters at 12 Bessborough Gardens, London (since demolished under a redevelopment scheme).

## NEW HEADQUARTERS AND THE FUTURE

A search was made for alternative premises and an appeal for funds to make such a move possible was launched. Members and friends responded handsomely and eventually a suitable site was found. The building was derelict, but had potential. Eventually it was purchased, plans for its rebuilding were drawn up, and planning permission obtained, the plans modified to meet conditions imposed by various authorities (part of the site was occupied by a listed historic building and required special treatment), and builders chosen. The rebuilding was an anxious time for the Executive Secretary (who had to keep an eye on progress and watch out for any snags or misinterpretations of instructions by the builders, as well as carrying on with the routine business of the Society) and other staff, but eventually the construction was complete and by May 1979 the BIS had crossed the Thames to the South Bank and was installed in its new headquarters.[29]

The Society was no longer in rented accommodation, but in property it owned completely and it has been the Council's policy to continue to improve its Headquarters. We now had satisfactory offices, library, a Conference Hall, Council Room, a reception area for members, kitchens, storage space, etc. Equipment was updated (the BIS has now acquired word processors, a modern plain paper copier, and other advanced machines to expedite its work) and we are today considering further expansion.

**Figure 2** Present BIS Headquarters (1979) at 27/29 South Lambeth Road, Vauxhall, London, SW8 1SZ.

Back in April 1974, the first of a series of "red-cover" issues of the *Journal* had been published; these dealt with Interstellar Studies--a subject in which the BIS had a long-standing interest. "Red-cover" issues have continued to appear at frequent intervals. A group of BIS members has also been working on a starship study since 1972 and this culminated in a very successful book on Project Daedalus, as it became known.[30]

Other "special-color" issues of the *Journal* followed, each dealing with a particular topic in astronautics (orbital dynamics, space science, astronautics history, etc.) and each with its own "Series Editor". This enabled Professor Groves to relinquish the editorship of the *Journal*, and when Ken Gatland's professional commitments forced him to step down as Editor of *Spaceflight* in May 1981 (after a long and exhausting tenure of that office), Professor Groves was able to take his place. A third journal, *Space Education*, began in 1981, under the wing of *Spaceflight* initially, but now independent and edited by A. D. Farmer.

Recent books issued by the BIS include *High Road to the Moon* (which records many of the Society's original ideas and discussions on lunar exploration in the visionary drawings of the late R. A. Smith[31] and *The Eagle Has Wings* (which tells the story of major U.S. space projects from 1945 to 1975).[32]

The BIS is about to start its second half-century, and we have every confidence that even greater achievements lie ahead for our society and astronautics in general. We look forward to the establishment of large permanent spacestations, to a return to the Moon in greater strength to establish a lunar base, and (though perhaps not in the lifetimes of the authors of this paper) the first manned mission to Mars. The BIS hopes to continue to contribute ideas towards these projects. Indeed, with its interest in star travel, the Society sometimes gazes even further ahead. If the reader thinks that too distant to merit present discussion, let us remind him or her that two man-made vehicles have already left the Solar System. Who knows where they may go eventually?

**REFERENCES**

1. F. H. Winter, *Prelude to the Space Age: The Rocket Societies 1924-1940*, Smithsonian Institution, Washington, 1983.
2. L. E. Winick, *Spaceflight*, 1978, 20, p.162.
3. F. H. Winter, *Spaceflight*, 1977, 19, p.243.
4. *Spaceflight*, 1967, 9, pp.150, 201, 234, 264, 299.
5. G. V. E. Thompson, *Spaceflight*, 1979, 21, p.402.
6. A. V. Cleaver, *Spaceflight*, 1961, 3, p.169.
7. P. E. Cleator, *Rockets through Space or the Dawn of Interplanetary Travel*, Allen & Unwin, London, 1936.
8. P. E. Cleator, *J. Br Interplan Soc*, 1950, 9, p.49.
9. H. E. Ross, *J Br Interplan Soc*, 1950, 9, p.93.
10. H. E. Ross, *Spaceflight*, 1961, 3, p.164.
11. E. Burgess, Paper to 17th IAA Symposium on the History of Astronautics, 1983, IAA-83-287.
12. K. W. Gatland, A. E. Dixon, and A. M. Kunesch, *J Br Interplan Soc*, 1950, 9, p.155; 1951, 10, pp.97, 107, 115; 1953, 12, p.274.
13. *Spaceflight*, 1979, 21, p.227.
14. A. V. Cleaver, *J Br Interplan Soc*, 1950, 9, p.315.
15. L. R. Shepherd, *Spaceflight*, 1956, 1, p.159.
16. *Realities of Space Travel: Selected Papers of the British Interplanetary Society*, (Edited by L. J. Carter), Putnam, London, 1957.
17. W. S. Bainbridge, *The Spaceflight Revolution*, Krieger, New York, 1983.
18. *High Altitude and Satellite Rockets: A Symposium Held at Cranfield, England, 18th-20th July 1957*, Royal Aeronautical Society & British Interplanetary Society, London, 1958.
19. *J Br Interplan Soc*, 1960, 17, pp.278-327.
20. *Space Research and Technology: Proceedings of Symposia Sponsored by the British Interplanetary Society: Space Medicine Symposium: Rocket and Satellite Symposium: Space Navigation Symposium: Liquid Hydrogen Symposium*, (Edited by G. V. E. Thompson), Gordon & Breach, New York-London, 1962.
21. *Spaceflight Technology: Proceedings of the First Commonwealth Spaceflight Symposium Organized by the British Interplanetary Society, 1959*, (Edited by K. W. Gatland), Academic, London, 1960.

22. *Communications Satellites: Proceedings of a Symposium Organized by the British Interplanetary Society*, (Edited by L. J. Carter), Academic London, 1962.

23. *Materials in Space Technology*, (Edited by G. V. E. Thompson and K. W. Gatland), Iliffe, London, 1963.

24. *Rocket Propulsion Technology, Vol. 1. Proceedings of the First Rocket Propulsion Symposium. 1961*, (Edited by D. S. Carton), Plenum, London, 1961.

25. *Spaceflight Today*, (Edited by K. W. Gatland), Iliffe, London, 1963.

26. *Space Research and Technology. Vol. 1. The Space Environment*, (Edited by N. H. Langton), University of London Press, London, 1969.

27. *Space Research and Technology. Vol. 2. Rocket Propulsion*, (Edited by N. H. Langton), University of London Press, London, 1970.

28. *Teachers Handbook of Astronautics*, (Edited by S. W. Smith), British Interplanetary Society, London, 1963.

29. E. Waine. *Spaceflight*, 1979, $\underline{21}$, p.409.

30. *Project Daedalus: The Final Report of the BIS Starship Study* (Edited by A. R. Martin), British Interplanetary Society), London, 1978.

31. R. C. Parkinson and R. A. Smith, *High Road to the Moon*, British Interplanetary Society, London, 1979.

32. A. Wilson, *The Eagle Has Wings*, British Interplanetary Society, London, 1982.

AAS 91-285

## Chapter 5

## LIQUID PROPELLANT ROCKET DEVELOPMENT BY THE U.S. NAVY DURING WORLD WAR II: A MEMOIR[*]

### Robert C. Truax[†]

Substantially all development work done by the U.S. Navy during and immediately prior to World War II on liquid propellant rockets was conducted under the sponsorship of the Navy's Bureau of Aeronautics. The work fell into two classes - that done in-house (that is, by government personnel at naval facilities), and that done by private contractors.

The first work by the Navy was initiated in May of 1941 when the author was ordered to duty at the Bureau of Aeronautics. This assignment was a direct result of work done while a midshipman at the U.S. Naval Academy in 1937-1939.

**Figure 1** Midshipman R. C. Truax operating rocket thrust chamber using compressed air and gasoline as propellants, 1938.

---

[*] Presented at the Seventeenth History Symposium of the International Academy of Astronautics, Budapest, Hungary, 1983.

[†] President, Truax Engineering, Inc.; Fellow, AIAA.

Truax set up the first jet propulsion desk in the Ship Installations Division. This division was selected because the only interest in the beginning was in assisting the takeoff of large seaplanes and the rockets were visualized as a kind of catapult. In charge of the S.I. Division was Commander (later Rear Admiral) C. M. Bolster who, by his sympathetic understanding and encouragement of the work, greatly facilitated its progress.

The author's first task was to make a general survey of areas of jet propulsion of potential interest. The report submitted in June of 1941 recommended development of both air-breathing and rocket jet propulsion for manned aircraft, sounding rockets and missiles. Only the recommendation to begin a program for the development of a JATO (Jet Assisted Take Off) rocket for the PBY seaplane was implemented.

In charge of "Experiments and Developments" for the Bureau of Aeronautics was Commander (later Rear Admiral) L. C. Stevens, who controlled the allocation of funds for experimental purposes.

Following the author's recommendations, an in-house project for the development of the PBY JATO was set up at the U.S. Naval Engineering Experiment Station, Annapolis, Maryland, and the author was sent there as officer in charge, turning over the Bureau desk to Lt. C. F. Fisher (USNR).

Initial funding for the project was $55,000, twenty-five thousand of which went for the construction of special test facilities and the remainder for design, development and testing.

The first personnel assigned to the project included the author as Officer in Charge, Ensign R. C. Stiff, Ensign J. P. Patton, Ensign W. Schubert, and Mr. Robertson Youngquist.

Preliminary analysis of the problem indicated that a thrust of approximately 3000 pounds (13.kN) for a period of about 35 seconds would be required. Very early in the course of development, the major design characteristics were selected. For logistic reasons, the propellants chosen were red fuming nitric acid as oxidizer, and gasoline as fuel. The burning time required dictated the use of a regeneratively cooled thrust chamber, and the simplicity and reliability desired indicated that a compressed gas propellant feeding system would probably be the most desirable. In regard to the latter point, the advantages of generating the pressurizing gases from chemicals carried within the rocket were recognized and development work on suitable systems initiated.

Since at that time design data on liquid propellant rocket engines were almost non-existent, a program of small-scale tests was carried out to determine injector (atomizer) criteria, combustion chamber volume requirements, and the heat transfer rate to the combustion chamber walls. Studied also were means for igniting and maintaining combustion with the selected propellants.

**Figure 2** Project personnel with Aerojet Founders von Kármán and Forman. Front row left to right: Parker, House, Schubert, Patton. Middle row: Stiff, Forman, von Kármán, Fisher, Truax, Youngquist. Back row: Row, Gore, Hall, Haughton, Phipps.

In order to arrive at an optimum injector design without introducing additional complications, the program called for open air tests of various injector configurations. This procedure spread consternation among the project personnel when tests disclosed that the combination of nitric acid and gasoline could not be ignited, even with a blowtorch, and that combustion could be obtained only after the propellants had run together on the ground and remained there, with the gasoline burning in air for some time. After numerous failures, we were ready to abandon the propellant combination when it was suggested that high pressure and confinement might be prerequisites for proper combustion. With this thought in mind, the decision was made to "take the bull by the horns" and try a full pressure test of a complete rocket thrust chamber. The test was a complete success and excellent combustion was obtained. All further tests were made with the injectors installed in suitable test chambers. However, since enclosing the injector prevented visual observation of the initial flame pattern, and thus comparing reaction volumes, a new method of evaluating injector designs became necessary. A suitable criterion was set up by Mr. Youngquist, the parameter $Ap/w$, now almost universally called $c^*$. This parameter is the area of the nozzle throat (A) multiplied by the combustion chamber pressure (P) and divided by the mass rate of flow of propellants (W). It is numerically equal to the velocity of the gases at the throat and is an indication of the gas temperature in the combustion chamber and hence the combustion efficiency of the injector when used with a given chamber.

War was declared shortly after the first successful small-scale test, and under renewed pressure, the staff and budget of the project were rapidly expanded. The experimental program was accelerated accordingly. In late spring of 1942 design of the full-size motor was started. Since two units were to be used to give the 3,000 pounds (13.4 kN) thrust, each thrust chamber was required to deliver 1,500 pounds (6.7 kN) thrust. The 1,500-pound thrust chamber constructed was the largest built in the United States up to that time. The first firing took place in June. A lag in ignition by the powder squib igniter permitted a dangerous quantity of propellants to accumulate in the chamber. The result was an explosion that destroyed the entire stand.

While the motor program had been proceeding, Ensign Stiff, in charge of the chemical gas generator program, had discovered that several chemicals ignite spontaneously with nitric acid; among these, aniline was considered the most practical. This information was communicated to the rocket group at the California Institute of Technology, which was working on a similar program for the Air Corps, and also using nitric acid and gasoline as propellants. Although ignition was readily effected on small-scale tests by pyrotechnic means, ignition of large-scale motors was more difficult, and ignition lag meant, usually, a disastrous explosion. In search for an alternate method, this group proceeded to use aniline in place of gasoline as the fuel, with excellent success. In spite of the logistic argument against its use, pressure of time forced the project also to turn to aniline for a solution to the problem of ignition.

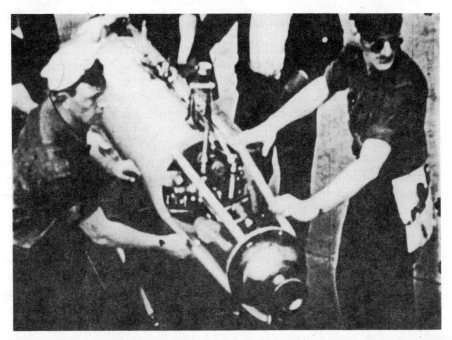

**Figure 3** DU-1 droppable JATO used on PBY seaplane.

The second thrust chamber, using aniline as fuel, was successful, and was used with only minor refinements in the final unit.

**Figure 4** PBY seaplane taking off with extra boost from two DU-1s.

This unit was designated the DU-1; droppable unit number one, for it was designed to be dropped by parachute after use. It was rated at 1,500 pounds (6.7 kN) thrust for 35 seconds. Weight empty was about 325 pounds (147 kg), and loaded about 655 pounds (297 kg). Two units were used, one suspended from the wing struts on either side. Takeoff tests were conducted in the spring of 1943, and were completed without incident. The average reduction in takeoff distance was 60%.

The flight test crew of the project found no lack of work to do. Very early in the history of the organization, Dr. R. H. Goddard offered his services to the Navy and was awarded a contract to develop an assisted takeoff unit based on liquid oxygen and gasoline propellants. Dr. Goddard brought this group from New Mexico, where he had been working for several years under the sponsorship of the Guggenheim Foundation, and set up shop at Annapolis alongside the Navy project. His unit was the first to be pronounced ready for flight tests, and it was accordingly installed in the PBY-2. This unit was not designed to be dropped, but was built into the after end of the airplane. Bad luck dogged the footsteps of Dr. Goddard, and, after a fire that severely damaged the airplane, further flight tests were discontinued.

After the DU-1 tests, a new test airplane arrived, a Martin PBM, and into this a JATO unit was built under a Navy contract by Reaction Motors, Inc. of Pompton Plains, N.J. This installation delivered 3,000 pounds (13.4 kN) thrust for 60 seconds using liquid oxygen, gasoline, and water as propellants. Tests with this unit were highly successful.

In two concurrent projects, the Bureau of Aeronautics contracted with Aerojet Engineering Corp. to develop a refined version of the DU-1 and also a pump-fed JATO to be permanently installed in the hull of a PB2Y3, a four-engined seaplane. This contract followed the successful flight demonstration of the pressure-fed experimental version on the same airplane.

**Figure 5** Gas generator using nitric acid and aniline as propellants built and tested by Truax in 1943.

Truax and Stiff (now Lt. and Lt. jg. respectively) were ordered to duty at Aerojet to monitor the two programs. Truax took the pump project and Stiff the droppable JATO. Eventually approximately 100 of these latter units were built.

**Figure 6** Gorgon rocket engines using nitric acid and aniline as propellants.

These Aerojet rockets, designated 38ALDW1500, were flight-tested by the Annapolis Project flight test crew, under the supervision of Captain J. L. Gore, USMC, Chief Test Pilot. They performed very well. Of all the liquid propellant JATO developed, the 39ALDW1500s came the nearest to service use. It was used to a minor extent by the U.S. Coast Guard Station at San Diego, California, for offshore rescue work.

The pump-fed JATO also used nitric acid and aniline propellants and had a thrust of 6,000 pounds (26.7 kN) in three 2,000-pound (8.9 kN) thrust chambers. The pumps were driven by a small reciprocating engine. This JATO never reached the flight-test stage. Much of the difficulty with the pump-fed unit stemmed from problems with the gearbox necessary to marry the reciprocating engine to the pumps. So the author, following Goddard's lead, but unaware of the latter's pump work, suggested that direct drive turbines be substituted. He designed, built and tested a small gas generator to drive such a turbine.

**Figure 7** Gorgon IIA Air-to-Air Missile (front view).

Data on this gas generator was turned over to Aerojet and led to the first really successful pump-fed engine - a 6,000 pound (26.7 kN) thrust engine ordered for a rocket propelled manned airplane that never flew.

Meanwhile, Truax came up with the idea of driving the turbine with the main propulsive jet and using a partial admission turbine with air cooling to solve the temperature problem. A turbine was built and successfully demonstrated while Truax was at Aerojet. Shortly after this, Truax was ordered back to Annapolis to resume the reins of the projects at the Engineering Experiment Station (EES).

By this time, the liquid JATO work was tapering off, but the project carried on a great deal of work in propellant research. New applications also arose.

**Figure 8**  Fairchild LARK Surface-to-Air Missile.

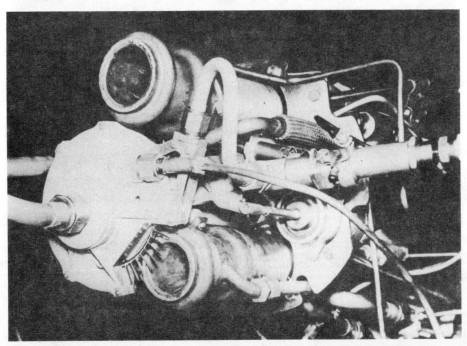

**Figure 9**  Pump-fed LARK engine showing booster thrust chambers and unique turbine driven by sustainer exhaust.

News came to the group at Annapolis that the Naval Aircraft Factory was toying with the idea of using a small, fast, radio-controlled airplane to carry an explosive charge that would destroy enemy bombers. The power plant for this missile was to be a small turbojet, but the engineers at the Aircraft Factory were rather appalled by the cost of this engine for an expendable airplane. As a result, the EES Project team was called upon to design a rocket engine delivering 350 pounds thrust for a period of about two minutes. This power plant was developed in the short space of 45 days, primarily through the intensive efforts of the project engineer, Lieut. William Schubert. Test quantities were manufactured by Reaction Motors. Test flights of the Gorgon (as the missile was called) were successful as far as stable flights were concerned. The rocket engine drove the Gorgon to speeds in excess of 500 mph (224 m/s) an unheard of speed for the time. Propellants used were mixed nitric/sulfuric acids and monoethylaniline.

**Figure 10** High-altitude cooling tests of air-cooled "blast" turbine LARK engine.

When Truax returned from Aerojet, he initiated considerable work in the direct exhaust-driven turbo pump (sometimes called the blast turbine). An experimental engine was operated using a Gorgon thrust chamber and a blast turbo pump.

At about this point, the Bureau of Aeronautics laid down requirements for a surface-to-air missile to counter the Japanese Kamikaza threat, and work was started on two versions of the LARK, as it was called. One contract was let to Fairchild Airplane & Engine Co. and one to Convair. The engines for both were based on the EES-developed Gorgon thrust chamber. Convair chose to stay with a gas pressure-fed system, and Fairchild opted to begin with a pressure-fed system but to switch later to a version using Truax's blast turbine. Contracts were let with Reaction Motors, Inc. for the thrust chambers and to Eclipse-Pioneer for the pumps.

Difficulties with the turbine delayed the pump engine, but several LARK eventually flew with this type of power plant.

The LARK was the first U.S. anti-aircraft missile to hit a drone target, but it never went into service.

During this period, Dr. Goddard also turned his attention to missile engines. Very reluctantly, he finally switched from his favorite propellants, liquid oxygen and gasoline, to nitric acid and aniline. He experimented with numerous methods for driving pumps, including a Pelton Wheel type of blast turbine, and a rotating combustion chamber.

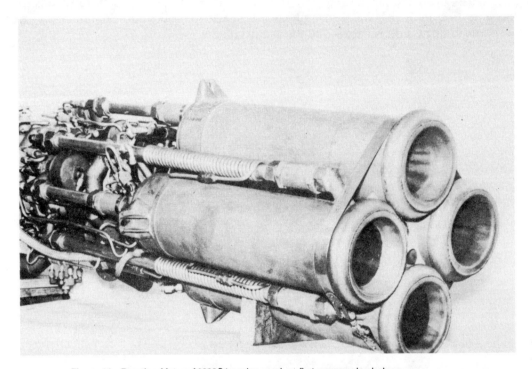

**Figure 11** Reaction Motors A6000C4 engine used on first supersonic airplane.

Later in the World War II period, the Bureau of Aeronautics initiated development of a 6,000-pound (26.7 kN) thrust engine for manned aircraft, the Reaction Motors A6000C4. A pressure-fed version of this engine was used in the Bell X-1, the first airplane to exceed the speed of sound. A pump-fed edition was installed in the Douglas D-558. Later versions of this engine were also used in early tests of the X-15 research airplane.

During its four-year history, the Annapolis Rocket Project grew to a staff of about 20 officer-engineers, some 50 enlisted men and about the same number of civilian technicians.

**Figure 12** Officer-Engineer staff of Annapolis Project, 1944. Front row, left to right: Simmers, Youngquist, Patton, Hart, Truax, Speer, Greenberg, Sumner, Fearn. Back row: Frazee, Gilpin, Gray, Leafgreen, Hull, Omang, Edelman, Cooley, Nyborg, Gray.

The end of the war saw the incorporation of the Annapolis Project into the Naval Air Missile Test Center, Pt. Mugu, and the cessation of rocket work at the Engineering Experiment Station. The personnel dispersed to other locations, and many continued to make important contributions to the field of rocket propulsion through long and illustrious careers.

AAS 91-286

## Chapter 6

# SOME VIGNETTES FROM AN EARLY ROCKETEER'S DIARY: A MEMOIR[*]

**Bernard Smith
as told to Frederick I. Ordway, III[†]**

### THE EARLY NINETEEN THIRTIES

When the Great Depression of 1929 appeared, those who had been living in real or imagined affluence suffered a catastrophic change in lifestyle. The rest of us hardly noticed any difference. But, years of staying alive by doing odd jobs wore thin; everything seemed likely to remain at subsistence levels forever. Having much time to spare I used the free New York City libraries and museums for all they were worth to store in my memory as odd a collection of pseudo-scientific smatterings as ever led a naive youth into believing a better world existed.

Where was it? Certainly not on this planet. At this point we come to the psychological counterpart of the continental divide and the watershed effect: an announcement that the American Interplanetary Society would hold one of its meetings in the American Museum of Natural History. There it was: the means for going to a better world wherever it existed. And that is how I came to join the ranks of yesterday's pioneers of the future.

The American Interplanetary Society was pulled together by a number of science fiction writers who believed that interplanetary travel was imminent, needing hardly more than a few mechanics to put things together. They did indeed have a mechanic in the body of H. Franklin Pierce who was assisted by perhaps a dozen members dedicated to the task of determining how Pierce could do the work of thirteen men. When events did not proceed fast enough, the members entertained the notion of changing the organization's name to the American Rocket Society (ARS), since interplanetary travel might take awhile. I came on the scene at about this time, when G. Edward Pendray took over the leadership. He had returned from a tour of European interplanetary societies, had put the American Society on a more rational basis, and had helped to run some furtive static tests of liquid fuel motors patterned after German models.

---

[*] Presented at the Seventeenth History Symposium of the International Academy of Astronautics, Budapest, Hungary, 1983.

[†] Alabama Space and Rocket Center, Huntsville, Alabama, U.S.A.

I attended one meeting late in 1932 where much of the above was reviewed, thrilled to overflowing with what was going on but too timid to voice all the ideas popping into my head. When it was over, I caught Pendray's eye and suggested a way of salvaging the components of a rocket damaged in earlier static tests. Ed Pendray was never one to overlook any opportunities. One thing led to another and before I knew it the components were in my hands to be reborn Phoenix-like from the old ashes. Ah, the exhilaration of those days! They will never be recaptured. We were just a few optimistic individuals attempting to do on a shoestring what no one today seems able to begin without enormous subsidies and thousands of trained technicians.

And how was I prepared to undertake this grandiose project of building ARS No. 2? In the best possible way. I had a grammar school diploma, I had done some locksmithing, tinsmithing, and iron mongering. I had won a prize for sculpture at Cooper Union in New York City and knew something about using the Pythagorean Theorem as well as solving a quadratic. Not until later was Pendray aware of this splendid background--otherwise he may have had second thoughts about the whole assignment. When he realized what he was into, he wisely decided to become my mentor; and when life became too tough for me, he often created some work around his home in Crestwood where I could live and earn some money.

The best description of the rocket's construction was to say that it was "accumulated." The cowl over the motor was carefully and accurately fashioned from a discarded aluminum coffee pot; the turn-on valves were gas-cocks liberated from an old cooking range; the fins were pieces of balsa wood begged from someone who had some scraps. And so it went, looking more like a rocket from week to week. There is not much purpose is going to more design detail which is covered elsewhere. In the light of today's knowledge much of the design approach was erroneous. Putting the rocket nozzle ahead of the C. G. did nothing to improve stability, nor did placing the fuel orifice near the nozzle throat produce an aspirating effect. Fortunately, these mistakes were not catastrophic.

Finally came the great day at Great Kill, Staten Island. Pendray saw to it that everything was legal so he could invite the press and the newsreel photographers. I believe ARS members John Shesta and Carl Ahrens were responsible for the launcher. They set it up pointing a bit to seaward to ensure some remoteness to the rocket's return. Pendray and I were responsible for mounting and loading the rocket and I was honored with the privilege of turning on the fuel cocks.

The firing procedure was as follows: After the gasoline and liquid oxygen tanks were charged and closed, a stick was slid into a horizontal slot cut into the cock handles. The outer end of the stick was fastened to a lanyard that stretched to the firing pit. I had designed the handle to hold the stick until the lanyard had turned the cocks a quarter turn, whereupon the stick would fall away. With everything in readiness a gasoline-soaked rag placed under the nozzle was ignited and everyone ran to the pit to watch me pull the lanyard. But the angle of pull was not quite right and the stick fell away before the cock was turned. My professional reputation being at stake I knew what I had to do. I ran back to the launcher, replaced the stick, reignited the rag, payed out the lanyard in the right direction and pulled it, thereby becoming on 14 May 1933, a day before my 23rd birthday, the first lad in America to publicly launch a liquid fuel rocket.

Judging by the criteria exercised decades later at Cape Canaveral, wherein any rocket that rose at least six inches off the pad was pronounced a success, this one was a resounding achievement. It rose a couple of hundred feet before it exploded. A man in a rowboat offshore saw it fall and retrieved it for us in triumph. Good thing the fins were of balsa; otherwise it would be at the bottom of Lower New York Bay to this very day.

During the post mortem it was pointed out that I could have gone up instead of the rocket. In helping Pendray with the Lox, both he and I were steeped in the concentrated oxygen vapors which had permeated our clothing and therefore I was in more jeopardy than I knew when I replaced the stick near the burning rag. I explained my action by saying that I dreaded doing a lick more work on that rocket and wanted to see it disappear off into space so I would work on the next one. Ah, yes, the next one. . . .there is always the next one to light up one's eyes.

Next, came three rockets proposed by teams. Pendray and I comprised one of the teams. We had a great time jockeying for position with the others; yet despite the spirit of competition, we helped each other on every occasion. Shesta, who headed another team, often machined experimental parts for me. And Nathan Carver, yet another competitor, made available to me work space in a loft he had rented. However, getting started was difficult for all of us and more than a year passed before we had anything to show for our troubles.

The agonizingly slow progress of ARS Rocket No. 3 set my mind going in another direction, one that I hoped would be less inhibited by the outside world. I began to fiddle around with air breathers requiring less esoteric materials and processes. To the great cause of science I sacrificed a gasoline blow-torch which was useful on many an odd job. Nothing was too sacred to offer up to the advance of knowledge. I redesigned the components and mounted them on a vertical pivot such that the whole business could revolve in a horizontal plane (Figure 1). After some adjustments I was able to make that machine turn at a fair slip, spewing hot flame in all directions.

Pleased with myself, I offered to demonstrate my "blow-torch" motor the next time the Society met at Pendray's home in Crestwood. On the appointed day we gathered in his garage around my machine as I prepared to prime it for firing. This consisted of soaking the nozzle and preheating coils with copious amounts of gasoline and lighting up. During this operation I could detect the quiet exit of certain science fiction writers who will remain unnamed. But Pendray was steadfast, even when the burning gasoline dripped over his garage floor. When the motor settled down to a steady state everyone returned to see the fiery display and to conclude that the device did not have enough thrust to lift its own weight. Ever after Pendray proclaimed this to be the first firing of an "Athodyd" (ramjet). But no question about it: I was a menace to New York from the Bronx to Staten Island!

Shesta's rocket, ARS No. 4, was finished first and was fired in September of 1934 at Great Kills. It had four nozzles up front pulling the tanks between them, one behind the other. It went up, burned out a nozzle and looped back into the bay. Mine, ARS No. 3, was readied for firing in the same place later that month. The firmed-up design now consisted of a blast chamber leading into a long nozzle about which was wrapped, consecutively and concentrically, a gasoline tank, a nitrogen pressure tank, and a Lox tank. I reasoned that a form of regenerative heating and cooling would take place through the conducting walls. The hot nozzle would keep the gasoline from being frozen into mush by the Lox and the cold gasoline would keep the nozzle from melting; something like boiling water in a paper bag over a flame, the flame keeping the bag from becoming water-soaked and the water keeping the flame from burning the paper. A beautiful theory, but we never found out how it worked; all that aluminum acting as a heat sink was our undoing.

**Figure 1** Bernard Smith preparing ARS Rocket No. 3 in his basement apartment in New York City.

No one, least of all I, foresaw the difficulty of filling the oxygen tank on that frustrating day in September 1934. The fill hole was simply too small. As fast as Pendray poured Lox into it the boiled-off gas came out. It was impossible to cool the uninsulated tank fast enough to hold liquid oxygen. The test had to be aborted to everybody's disappointment. I don't know why I didn't go back to the drawing board when in retrospect it would have been logical to cut a larger fill hole and add a bit of blanket around the oxygen tank, to be left behind at firing. The first and only test of that model was destined to be its last. In 1939, when I returned to the scene of the crime, I saw it on display at the New York World's Fair. The rocket and I had made our peace.

The failure of ARS No. 3 also failed to cure my rashness; in fact, it allowed failure to go to my head. My thoughts now centered on a scheme for fastening to the soles of my feet a low-thrust rocket motor connected through flexible hoses to fuel tanks held in my hands. I was convinced that the natural reactions of my feet would keep me erect if I should tilt in any direction, and hovering could be accomplished with the greatest of ease. I abandoned this notion before anyone could ridicule me when I realized that the "hot foot" motor would never become popular; nobody likes to stand on an overheated platform throwing all that dust. Twenty years later, William B. McLean, Technical Director at the Naval Ordnance Test Station, used a "house-broke" reacting platform to show that the correct foot response did indeed occur naturally.

The economic low point in New York occurred for me about six months after the ARS rocket No. 3 test. And so in the Spring of 1935, I seized an invitation by my uncle to come west and seek my fortune. I left to John Shesta in New York a graphite motor machined out of a giant furnace electrode. He tested it after I had gone and notified me that it blew up immediately after ignition. Years later, in California, I learned to make one that didn't blow up. So ended my early attempts to leave the planet from New York City.

Yet, as I moved westward from state to state on my way to California, schemes for accomplishing that end persisted. I dreamed of a track running up the side of an Andes mountain on which a tremendous booster would accelerate a space ship to March 3, at 20,000 feet (6,100 meters). Near the very peak the last two stages would be released and, with the Earth's easting, bring the ship to satellite velocity in the proper orbit. Long afterward, I read a Russian report describing a similar proposal. But I would not be in charge of a real satellite project until almost a quarter century had passed.

## VALHALLA OF THE AMATEURS (1935-1948)

Entering the enormous overgrown village of Los Angeles in 1935 by way of Valley Boulevard, I came at last to the heartland of the Southern California Chamber of Commerce. The scenes I saw have been compared to the finest tourist haunts of the Holy Land, Egypt, the Riviera, Greece, Bermuda, and sometimes with California itself. Actually there is no comparison. Only at one point could you believe you were entering a foreign country - the port of entry at the California border where your belongings were searched by customs officials. However, if you came by air your belongings were untouched and thereby you lost even that illusion.

Shortly after arriving at my uncle's home in Los Angeles, I took evening classes in welding to bring my blacksmithing up to modern times, and in mining, to help me discover gold and oil. The first paid off handsomely in providing the means to a modest livelihood for the next twelve years. There was no payoff to the second except for a little gold dust laboriously panned in the San Gabriel Canyon, soon lost in debauchery. The real gold was in the welding and once I gathered enough skill, I was never out of work thereafter. Near the end of the first year in California, my progress in welding opened doors to design experiences and shop practices that served me well for the rest of my life.

Meanwhile rockets were still on my mind. John Shesta had written to request some conductivity tests on various metals if at all possible. He was not sure that conductivity through metals constituted the overriding characteristic; chemically attacked surfaces at high temperatures could change the picture. Obviously he was working with James H. Wyld on a regenerative motor wherein the liquid fuel itself would be used to cool the combustion chamber, although he did not say so. However, Shesta had helped me in the past and I was determined to help him now.

I designed and built a fixture that allowed an oxy-acetylene flame to pass inside a specimen tube while a jacket of flowing water bathed its outside surface. The tubes I collected were thin-walled, of the same diameter and wall thickness, mostly of copper, aluminum, steel and their alloys. The measurements were very simple. With the flame passing through the tube, a fixed quantity of water was run by gravity through the jacket and collected in a container at the outlet. A thermometer in the collecting container told the story. To check on the flow rate, each run was timed and, not surprisingly, was found to be the same. Much more to my surprise I found the heat transfer through the tubes differed by very little, probably attributable to the thinness of the tube walls.

I concluded that under the conditions stainless steel was about as good as any and probably the most practical. I passed the test results along to Shesta but never learned how it was used, if at all. Many years later, when I had occasion to visit Reaction Motors in New Jersey, where the regenerative motor was being manufactured for the Defense Department, he briefed me on its design. It suddenly stuck me that I had conducted the original tests incorrectly: The cooling water should have passed inside the tube and the hot blast outside, against half the wall. But I noted that the tubes lining the inside of his blast chamber looked suspiciously like stainless steel. Perhaps my small contribution was not in vain after all.

Sometime during the year of 1939, Robert Gordon, packing a membership card in the American Rocket Society, showed up at my front door to form a California Rocket Society. He had written to the American Rocket Society about his idea and they quickly assured him only one man was qualified to form it, namely me. Pretty soon even I came to believe it a grievous mistake. Nevertheless, we proceeded onward and upward and with a few announcements in the newspapers collected an increasing membership at monthly meetings, convened in the Los Angeles Museum near Exposition Park.

As president of the new rocket society, I supposed it was my responsibility to come up with rocket designs and to organize tests. And so I put together a number of small projects to keep the members constructively occupied. These were basically schemes for stabilizing and streamlining common black powder rockets and though they did little to advance rocketry, they did wonders for morale insofar as they delivered fire and smoke and other visible tokens of action such as the successful and gentle lowering of payloads by parachute (Figure 2). However, I suspected that this fiddling around would soon wear thin; the burning problem was reaching great heights, payloads had a habit of coming down by themselves anyway.

**Figure 2** Members of the California Rocket Society preparing to fire a rocket to test a parachute-release device to recover the vehicle without damage, ca 1943.

We conducted our tests in the Arroyo Seco, a parched riverbed near Pasadena. So sure were we of success, we invariably let the press know about it. It was after one of these sessions that I came out with my most profound statement, one that is destined to ensure my undying fame. "Will these things ever be used in warfare?" one reporter asked. "Not at all", I replied knowingly. "Rockets were last used for warfare during the 19th Century, but they have since been superseded by projectiles fired from rifled guns, which are far more accurate. However, the rocket has excellent potential as a scientific instrument, for probing high altitudes and as an aid in meteorology. Only in that sense does it have military value." I think those reporters were pretty ignorant to be taken in by such a bald-faced pacifist statement when only a short distance away Franklin Malina and his crew at the California Institute of Technology were successfully developing military rockets for the U.S. Defense Department. To ease the reader's mind, I must state immediately that this was not the first nor the last time I was dead wrong as a result of ignorance in my own field.

Overstimulated by the conflicts in all the walks of life about me, I came at last to a rocket concept that satisfied the peculiar requirements of an amateur rocket society: the solid-fluid reaction motor, I reasoned, why not pass an oxidizing fluid over a reducing solid? A dense high-temperature solid without any oxidant admixture could only burn at its surface and therefore could never produce a large sudden explosive mixture, nor would there be a problem of pressure balance between fluid co-fuels. A single plumbing line could meter the fluid; thus the system was both safe and controllable. With a few of the more enterprising colleagues in the society we promptly pursued this notion.

The cheapest, most practical combination turned out to be carbon and oxygen, although the specific impulse promised to be only about that of black powder. More esoteric fuels were feasible but not within the society's scope of operations. I picked over a bunch of broken furnace electrodes in the Southern Pacific salvage yards and found exactly what I needed, a solid cylinder of graphite and another of pure massive carbon without voids. The graphite was machined to form a liner and nozzle inside a heavy-walled and capped steel tube. The carbon rod was fluted to increase burning area and machined to fit freely inside the liner. Oxygen would be fed through the cap as a gas; I reasoned there was no need yet to complicate tests with liquid oxygen since once the motor was in operation the liquid would gasify before it flowed over the hot plug.

We made everything ourselves including the test stand for measuring thrust. The remaining problem was ignition. I thought about packing black powder between the plug and the liner and inserting a piece of fuse cord. But this would be betraying my best friend, Gavin Calceron, who graciously allowed us the use of his backyard as a test site on my assurance that nothing would explode. Fortunately, the motor was to be tested upside down, which made it easy to insert a coated electrode through the nozzle against the tip of the carbon plug. A heavy current supplied by a bank of car batteries could bring the plug up to the ignition temperature.

By courtesy of the Fruehauf Trailer Co., I borrowed an oxygen tank, gauges and pressure hoses for the big event. With everything in readiness and a movie camera grinding away I inserted the electrode into the motor, waited a few seconds until I could spy a glow at the tip of the plug and withdrew with a signal to Gordon to turn on the oxygen. It worked with a roar! We could turn it on and off and regulate thrust at will! A great success but not perfect, for we soon found free oxygen in the exhaust which indicated insufficient burning area in the plug. Increasing the burning area then became the next problem, never really solved because the society was overtaken by other events.

At about this time the facts of life caught up with me. I was being bypassed by well-financed rocket activities in which I could not participate without more advanced training. Moreover, the society's members were as poor as I was and too many of them had interests bordering on the occult rather than on realities. The jig was pretty much up for rocket amateurs.

It didn't take much courage to decide that the only logical thing left to do was to get myself a different union card - a college degree. But no college would admit me without a high school diploma except Reed College, Portland, Oregon and then only after I had successfully passed about 18 hours of written examination. So I took the exams and one fateful day received notice that I was entered as a *special student*. In short order, I sold my house, liquidated all holdings, packed my wife, daughter and all other portable belongings into and onto my trusty Studebaker Champion and headed for Portland, in the late summer of 1944, to become a freshman in my 34th year of life. And so for the next four years I came under the wing of Dr. A. A. Knowlton, head of the Physics Department of Reed College; he approved my entry there and realized that it was his responsibility to get me through.

## THE MAKING OF A PROFESSIONAL NEOPHYTE (1948-1959)

In my junior year, Knowlton was visited by a Dr. Loeb, a friend who was, I discovered later, a highly qualified scientist as well as a naval officer. He came to recruit scientists for the Naval Ordnance Test Station (NOTS), in California, a rocket and missile research, development and test center just established by the Navy in the middle of the Mojave Desert. Knowlton felt that because of my previous rocket experience Loeb should interview me for employment and arranged a meeting. I think Loeb was taken aback when he saw a gray-headed junior, but with recruitment not going too well he offered me a position as soon as I graduated. I promptly dismissed the whole idea, finding the thought of working for the military not too inviting. I didn't want to develop weapons; I wanted to develop high-altitude scientific probes. But fate had already ordained that I would get to the probes through the weapons. Knowlton bided his time.

Only because I was desperate for work did I apply for it at NOTS in 1948. At that time they were ready to take almost anyone with a diploma. Thus did two desperations join hands to form one contentment. And thus we see that oftentimes good decisions can be made during periods of quiet desperation. Wherever he is, Dr. Knowlton must be smiling over the unobtrusive way he steered me into channels that shaped the rest of my career. He was great, I miss him. I leave to the reader one of A. A. Knowlton's more profound statements. "The practicing physicist tries to reconcile concepts with precepts. He doesn't prattle too much about 'truth'".

I view my quarter century of research and development in the Navy as the most exhilarating time in my life. It began with instructions from a sailor at the landlocked Naval Ordnance Test Station to go "through the bulkhead, up the ladder to the next deck" for formal processing as a U.S. Civil Service rating GS-7 and ended a quarter century later with my retirement as a GS-18 with the creation of the annual Bernard Smith Award, for Navy professionals who have rendered their services in the face of great odds. The path from one to the other is strewn with projects covering the development of squibs, igniters, rockets, warheads, projectiles, torpedoes, guided missiles, sensors, fire control, launchers and guns--some of each for surface ships, for aircraft, for submarines, and many undertaken simultaneously. By the time I faded away hardly a combatant vessel in the Navy escaped being furnished with some equipment I invented, initiated, developed or managed to deployment. To screw up the density of activity yet another turn, throw in a satellite and a deep-space probe smack in the middle of my career and you pack into it enough excitement to last a few lifetimes.

A squib is a small device that converts electric energy into a thermal shock. This sets off a pyrotechnic chain serving purposes as far apart as blasting ore out of a mountain or launching a space rocket. My first job at NOTS was to replace the squib furnished reluctantly by the DuPont Company, which refused to specify what it was selling to the Navy and, moreover, served a notice that it would not fill any more orders after a certain date. I never finished the job. Before I had gone very far I was assigned responsibility for five rocket igniters, each of which was supposed to contain those selfsame squibs. By the second year, assistance on an anti-tank, shaped-charge weapon and an anti-air rod warhead were added to my duties. By the end of the third year, I was developing a ship-launched, nuclear-headed monster nicknamed "Big Stupe," and by virtue of some human kindness I was relieved of my other responsibilities, but only for the six months allotted to the project.

Big Stupe turned out to be a two-stage rocket containing a couple of tons (1,800 kilograms) of propellant. It weighed five tons (4,500 kilograms), making it the largest rocket ever launched to that time at NOTS. To help me as much as possible, I was given one boss and two assistants. Again in the spirit of helpfulness, my boss locked the four of us behind a secret door and never permitted anything to come or go unless it was covered with a black shroud. These measures aroused the curiosity of all 4,000 employees on the base and soon the children in the government schools at China Lake were drawing pictures of the secret weapon "Big Stupe." Nevertheless, the project went along smartly because there were so few involved. Six firings, each with successful second-stage separation, were carried out. Within a week of the last firing, the complete report was delivered in Washington exactly on the final project day. But our hasty little project successfully and unintentionally proved that for those times it was far too cumbersome to be used aboard ship. Today we launch with greater ease much larger missiles from submarines.

My next assignments came in a bunch: the antisubmarine weapon, Weapon A, Weapon A subcaliber and Weapon B; the 5-inch (13 cm) flare rocket; the 2.75 (7 cm) subcaliber SCAR; the tank-launched Line Charge, and a few 5-inch spin-stabilized Barrage Rockets. This was enough to make me a Branch Head and accordingly I was promoted into the ranks of GS-13s (Figure 3).

**Figure 3** Smith (left) at work at NOTS.

These programs were mostly hand-me-downs started by others but left unfinished, in trouble, or ready to be cashiered. Now came something different from the best combination of a ballistician and a politician to be found in the old Navy's Bureau of Ordnance. Albert Wertheimer, who had responsibility for anti-submarine torpedoes would always arrive where the submarine had left if they had to swim all the way. He proposed hurling the torpedo to the last known submarine position with a rocket boost. The response was not even luke warm. Torpedoes were not designed to take the rocket acceleration or water entry at that speed. Moreover, what was the use of projecting the torpedo beyond the range of the ship's sonar? And the sonar people were asking, "What's the use of extending sonar beyond the range of the torpedo?" So Wertheimer, grimly determined to break this circular reasoning, had to prove his point by circumventing the torpedo people. He sent an assignment to NOTS entitled "Improvement of Ahead-Thrown Weapons," which covered a scheme to project a mobile mine out to 3,000 yards (2,700 meters). A mobile mine is a homing mine with self-contained propulsion, about as close to being a torpedo as one can get. This program kicked around for a while and finally landed in my lap.

I was immediately taken with the concept and decided we would do it in Big Stupe fashion: six firings within six months ending with the final report three days after the last firing. The mine chosen, Mark 24, was most appropriate, being already fitted with a parachute for air release that would reduce water impact. The rocket motor most nearly of the right impulse turned out to be an excess JATO bottle. About these two available components a design was quickly prepared and the necessary parts fabricated on the base. Sure enough, the project was completed within the self-allotted time and the report delivered in hand by me to Wertheimer. I found out much later that he was impressed, but at the time he never let on.

By the second year after the new program's inception, the Rocket Assisted Torpedo (RAT) was ready to be tested at sea. Then the fun began. No sooner had it made its first demonstrations and talk was concentrating on pilot product when a captain at a desk above Wertheimer became intensely interested in the program. "Why 3,000 yards (2,700 meters) range?" he asked, "Why not 5,000 (4,600 meters)?" Being a man of action he answered the question himself. "Make it 5,000 yards," he said.

And so, after a lapse of two more years, the Super RAT was ready for sea trials again and, would you believe it?, a still higher authority then asked, "Why 5,000 yards (4,600 meters) range? Why not 10,000? Let's make it 10,000 yards (9,100 meters)." I had visions of old man Smith starting all over again far into his advanced years on Super Duper RAT. But once again I met the challenge with a new design incorporating a large torpedo, a more powerful motor, a more ambitious launcher, and a more sophisticated everything else. At this point everyone was trying to get into the act. The concept that no one loved when Wertheimer was trying to sell it suddenly became the biggest "me too" bonanza in the world. His comment was most apt. "My enemies I can handle," he said, "but God help me from my friends." The weapon, now named "Asroc," had reached respectability.

Having served apprenticeship at every rung on the ladder from designer of small components to manager of all our organization's programs at one time or another, I eventually became head of the Weapons Development Department. I felt personally attached to every project, even to Sidewinder, the brainchild of William B. McLean, Technical Director of NOTS; that air-to-air rocket had components in it that I had designed. He spent more time with me than with the other department heads simply because headquarters had authorized an improved Sidewinder--a near impossible task--and he wished to keep his hand on it. However, McLean's intense interest in the details of the new Sidewinder design overflowed into all the other design work of the department, which made for a somewhat less than happy situation.

I reasoned that the best way to cope with the problem would be to initiate one or more programs that I could run according to my own preferences, while taking care of administrative problems (disliked by McLean) associated with the programs that occupied his attention. The opportunity arose when a Navy flier with memories of the Korean conflict came to me with the problem of trying to evade the enemy's gun-laying radar while attempting to drop a load of bombs on target. What could we do to blind that radar? It struck us both simultaneously that the radar beam itself was an excellent homing signal.

Forthwith, I put a team together to work on an air-launched, anti-radiation missile (ARM) and received instant support for it in Washington. The shoe was now on the other foot and McLean, to my surprise, didn't like it. He immediately generated competing ideas for achieving the same end and felt that I should be following his design notions, which I resisted. It was the first time I found McLean to be technically in error. To put it simply, he had so much trouble with D.C. (Direct Current) signal biases in Sidewinder circuitry that he felt the best solution for ARM was to spin the whole missile at a high rate and to convert all signals and responses to A.C. (Alternating Current). It was a disarmingly simple notion. But D.C. biases were no longer a severe problem in missile circuitry whereas spinning a homing missile added enormous and needless complexity to its design, presenting control problems never before encountered. I so advised McLean who remained unconvinced and I so advised the headquarters research and development chief, who was promptly convinced. ARM became the Shrike and remains in the arsenal.

The next big shambles in the sky took place a short while after Sputnik so startled the western world. We all watched in chagrin while the U.S. Navy struggled unsuccessfully time after time at Cape Canaveral to match that achievement. Unable to stand it any more, Mclean and Howard Wilcox, my former boss, traveled to Washington, D.C. to meet with Roy Johnson, head of the Department of Defense's Advance Research Projects Agency. They carried with them a proposal for putting into orbit a satellite in only four months at a cost of $300,000. Johnson accepted the proposal.

The idea was excellent. The satellite launcher's first stage was to be an F4 aircraft (a Navy fighter-bomber). At 40,000 feet altitude (12,200 meters) moving in the proper direction, inclination and speed, the F4 would release a 2,100 pound (955 kilogram) five-stage rocket. The ignition of each successive stage would be accomplished by timers and horizon detectors. The flight path in each case would be ballistic. The satellite payload would be an orbiting radio beeper. Analysis indicated the feasibility of the idea. Here was a potential, low-cost system that could serve less affluent nations that might wish to place into orbit small scientific payloads. Practically all the work involved could be accomplished at NOTS (Figures 4-6 and Table 1).

The schedule appeared impossible. Nevertheless, we almost made it. The first two attempts failed, and the third try in 1958 ended in one beep heard at the right time by a receiver stationed for the purpose at Christchurch in New Zealand; it was never heard again. By then, we were out of time and had exhausted our funds, including monies generously added by the Bureau. Wertheimer, who had suggested we put up a satellite years before Sputnik, was wonderfully supportive. And everyone was sorry. Nothing was left but to go back to Roy Johnson in Washington with our report of failure, and ask for additional money.

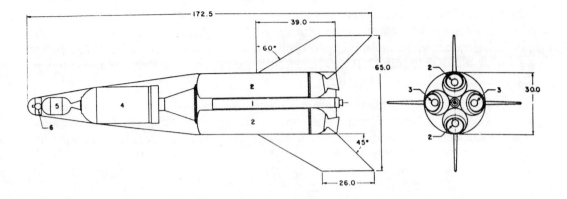

**Figure 4** Six-stage, air-launched NOTS satellite configuration.

**Table 1**
**DETAILS OF SIX STAGES OF NOTS SATELLITE**

| ITEM | WEIGHT LB | INITIAL WEIGHT OF STAGE LB | BURNT WEIGHT OF STAGE LB | TOTAL IMPULSE OF STAGE LB-SEC | BURNING TIME SEC | THRUST LB | NOZZLE EXIT AREA $IN^2$ | SPECIFIC IMPULSE $\frac{LB-SEC}{LB}$ |
|---|---|---|---|---|---|---|---|---|
| 1ST STAGE (ZUNI) MOTOR METAL PARTS PROPELLANT | 22 33 / 55 | 2104 | 2071 | 6800 | 1.0 | 6800 | NO CORRECTIONS | 200 |
| 2ND STAGE (2 HOTROCS) MOTOR METAL PARTS PROPELLANT | 120 600 / 720 | 2049 | 1449 | 138,000 | 4.86 | 28,396 | 190 | 230 |
| 3RD STAGE (2 HOTROCS) MOTOR METAL PARTS PROPELLANT FINS, STRUCTURE & FAIRING | 120 600 130 / 850 | 1449 | 849 | 138,000 | 4.86 | 28,396 | 190 | 230 |
| 4TH STAGE (ABL X-241) MOTOR METAL PARTS PROPELLANT FAIRING & TIMER | 56 376 10 / 442 | 479 | 103 | 97,930 | 36 | 2720 | VAC | 260.1 |
| 5TH STAGE (8-IN. JPN) MOTOR METAL PARTS PROPELLANT FAIRING | 6.0 26.9 0.5 / 33.4 | 37.0 | 10.1 | 6590 | 5.7 | 1155 | VAC | 245 |
| 6TH STAGE (3-IN. X-14) PAYLOAD MOTOR METAL PARTS TIMER PROPELLANT | 2.0 0.55 0.35 0.7 / 3.6 | 3.6 | 2.9 | 172 | 1.0 | 172 | VAC | 245 |

It was a hopeless mission to Washington; Johnson was irritated by the addition of our failure to many others that occurred during the early years of the U.S. space program. Support was denied our "NOTSnik" because it did not come out high enough in his dollars-per-pound-of-payload-in-orbit criterion. It was a sad day.

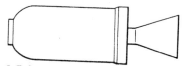

### ZUNI 1ST STAGE
LENGTH _____ 65.7 IN.
DIAMETER _____ 5.0 IN.
TOTAL WT. _____ 55 LB
PROPELLANT WT. _____ 34 LB
OPERATING PRESSURE _____ 200 PSI
BURNING TIME _____ 1.0 SEC
THRUST _____ 6,800 LB
SPECIFIC IMPULSE _____ 200 LB-SEC/LB
TOTAL IMPULSE _____ 6,800 LB-SEC
MOTOR PERFORMANCE INDEX __ 117

### ABL X241-4TH STAGE
LENGTH _____ 58.2
DIAMETER _____ 18.0 IN (NOM)
TOTAL WEIGHT _____ 432 LB
PROPELLANT WEIGHT _____ 376 LB
OPERATING PRESSURE _____ 200 PSI
BURNING TIME _____ 36 SEC
THRUST _____ 2720 LB
SPECIFIC IMPULSE _____ 260.1 LB-SEC/LB
TOTAL IMPULSE _____ 97,930 LB-SEC
MOTOR PERFORMANCE INDEX __ 225

### HOTROC
**2nd and 3rd Stage**
LENGTH _____ 71. IN.
DIAMETER _____ 11.65 IN.
TOTAL WEIGHT _____ 360 LB
PROPELLANT WEIGHT _____ 300 LB
OPERATING PRESSURE _____ 900 PSI
BURNING TIME _____ 4.96 SEC
THRUST _____ 14,200 LB
SPECIFIC IMPULSE _____ 230 LB-SEC/LB
TOTAL IMPULSE _____ 69,000 LB-SEC
MOTOR PERFORMANCE INDEX __ 192

### SPHERICAL 6th STAGE
LENGTH _____ 5.5 IN
DIAMETER _____ 3.0 IN
TOTAL WEIGHT _____ 1.25 LB
PROPELLANT WEIGHT _____ 0.7 LB
OPERATING PRESSURE _____ 1,500 PSI
BURNING TIME _____ 1.0 SEC
THRUST _____ 172 LB
SPECIFIC IMPULSE _____ 245 LB-SEC/LB
TOTAL IMPULSE _____ 172 LB-SEC
MOTOR PERFORMANCE INDEX __ 138

### EXTRUDED 5th STAGE
LENGTH _____ 18.6 IN.
DIAMETER _____ 8.0 IN.
TOTAL WEIGHT _____ 32.9 LB
PROPELLANT WEIGHT _____ 26.9 LB
OPERATING PRESSURE _____ 500 PSI
BURNING TIME _____ 5.7 SEC
THRUST _____ 1155 LB
SPECIFIC IMPULSE _____ 245 LB-SEC/LB
TOTAL IMPULSE _____ 6590 LB-SEC
MOTOR PERFORMANCE INDEX __ 200

**Figure 5**  Propulsion units of NOTS satellite.

**Figure 6**  Preparing the NOTS air-launched satellite for test.

Our satellite project was placed in cold storage to await a request from the Naval Research Laboratory for a 1,000-mile (1,600-kilometer) deep space probe required for the "HiHo" program. Our project was revived as Caleb, which employed an aircraft first stage as originally planned, successfully accomplished its task. My assistant, Charles Bernard (now director of land warfare in the Office of Secretary of Defense) mothered the project to its conclusion in 1959. So all was not in vain.

At about this time, after a continuous diet of crash programs for almost 12 years, I was sorely in need of a sabbatical leave; and, accordingly, the Navy created one for me at the Naval War College in Newport, Rhode Island. This assured for the Navy my service for another 14 years.

In leaving NOTS, I could not help but feel that the part of me born in liquid oxygen and gasoline in New York City had died in the Mojave Desert in California a quarter of a century later. The desire to leave Planet Earth had evaporated. Antarctica, the Sahara, and the oceans appeared by far more hospitable than the most attractive sites on other worlds or in interplanetary space. Nonetheless, I am most fortunate to have been among the few to enjoy the last chance to accomplish pioneering work in rocketry and astronautics as an individual experimenter, before all important programs became vast government-corporate enterprises.

## Chapter 7

## CONTRIBUTION OF THE ROMANIAN INVENTOR ALEXANDRU CHURCU TO THE DEVELOPMENT OF THEORETICAL AND PRACTICAL REACTIVE MOTION IN THE 19TH CENTURY[*]

### Florin Zăgănescu[†], Rodica Burlacu[‡] and I. M. Stefan[**]

The first Patent received by a Romanian inventor was obtained in 1827 by Petrache Poenaru, a student in Paris for an original fountain-pen, described as ". . . an unended portable pen, having its own supply of ink." Having the number 3208 and dated 25 May 1827, this patent was delivered by the French Bureau of Manufacturers.

Another patent was obtained also in the 19th Century by the another inventor, the Romanian Alexandru Churcu, (in Romanian, Ciurcu) aided by the Frenchman Just Buisson. It was a patent for a 'Jet Propulsion Engine' (the French original name was 'Propulseur à réaction), granted by the French Ministry of Commerce and Industry with the number 179001 and dated 12 October 1886.

Alexandru Churcu, one of the reactive motion pioneers, was born on 29 January 1854 in Sercaia village, the district of Făgăras, where his parents were exiled after the 1848 Revolution in the Romanian Province of Muntenia. The Churcu family came back to its native region only in 1856, when the exile ended. Alexandru was a practical and intelligent boy, with technical and cultural preoccupations. After studying law at the University in Vienna, Churcu returned to his country and founded in Bucharest some independent newspapers like *L' Independance Roumaine'* (The Romanian Independence). His opinions on justice and liberty and, of course, political ideas inherited from his father, led Churcu to criticize vehemently the unreasonable politics of the Romanian government headed by the chief of an old reactionary party.

---

[*] Presented at the Seventeenth History Symposium of the International Academy of Astronautics, Budapest, Hungary, 1983.

[†] University Professor, Scientific Secretary, Commission on astronautics of the Academy of the Socialist Republic of Romania.

[‡] Scientific researcher in chemistry; ICITPR.

[**] Writer; specialist in History of Science.

Using a contemporary method, a diabolical plan was carried out against Churcu, which was manifested by a sustained campaign to discredit him. The infamy having success, Churcu was exiled, on grounds of some unclear data concerning his Romanian citizenship on his passport.

In 1882, before his exile, Churcu conceived a very interesting and original kind of jet engine; he intended to use the reactive force of a jet exhausted from a cylindrical chamber where some special combustibles burned, to move by reaction various types of transport vehicles, especially an aerostat!

The shock of his expulsion was double, Churcu being a patriot very close to his people with its desire of liberty, independence and progress. Forced to go abroad, Churcu's choice was Paris, because he had in mind the possibility to get in touch with other Romanians having the same patriotic ideas. We may add he couldn't have forgotten that Paris was also the capital of new ideas and technical achievements about the real possibility for man to fly!

In Paris he was helped by a friend, the former Havas Agency correspondent in Bucharest, Just Buisson; he introduced A. Churcu to the cultural and aeronautical groups in Paris ...

## A NEW TECHNICAL SOLUTION IN REACTIVE MOTION AND ITS PROOF

In Paris, Churcu and Buisson were in touch with Edmond Blanc, His Excellency the Count of Hérisson, the inventors Gaston Tissandier, Emil Sarrau and Paul Vielle, the discoverer of smokeless powder.

Showing confidence toward Churcu's technical solution for the propulsion of any vehicle using reactive force, the two young inventors considered it to be useful for a lighter-than-air vehicle too. Churcu read passionately all the information concerning the aerostatic evolution. When the 1870 war ended, the French War Ministry nominated an aerostatic commission to promote the manufacture at Chalais-Meudon factory some dirigeable balloons with various dimentions. In this factory worked two specialists representative for this period: the officer-engineers Renard and de la Haye. Based on the old ideas of Henri Giffard (1885) and Dupuy de Lôme (1872) for mounting a propulsion system on a balloon, and having in view the success obtained by the inventor Gaston Tissandier with his 'electrical propelled balloon', director Renard and captain Krebs conceived, built, and exhibited at the Electricity Show in Paris their oval shape dirigible balloon, with a propeller driven by an electrical engine using an alkaline battery. It was very probable that Churcu and Buisson were present at the first 34-minute closed-circuit balloon-dirigible flight, successfully carried out on the 9th of August 1884 by Captain Krebs and Renard. Likewise, this 7.6 km long and perfectly controlled flight suggested, of course, the possibilities open to motorized dirigibles. It was impossible that Churcu and Buisson lost such an opportunity: They proposed to Gaston Tissandier to install on his balloon, instead of its electrical driver propeller, their own reactive engine, on which were mounted ducts and valves, for flight control.

But Tissandier was skeptical, and the Electricity Show organizers refused to take into consideration such an ineffective and dangerous, in their opinion, invention.

In their plans the two inventors were not discouraged, but calculated the possibilities that their engine could fly vertically, it was a wonderful anticipation in the period of 1880-1885 of the flight of heavier-than-air vehicles propelled by reactive engines. That would be a task for the future, accomplished by another renowned Romanian inventor, the scientist Henri Coanda (1880 - 1972), with the first-in-the-world reactive aircraft named 'Coanda-1910'.

Concerning the two inventors, they planned to prove the feasibility and efficiency of their invention on a vehicle. Another original idea was to install their reactive engine on a small boat. The first test with the world's first reactive boat took place on the 3rd of August 1886 on the Seine river, and it was characterized by Churcu himself as: "... The way on the Seine was very easy and our impression was splendid, as in a beautiful dream!" From the 3rd of August 1886 until the 16th of December 1886 they carried out other experiments, all successfully, and the jet engine was continuously refined by its inventors (Figure 1).

**Figure 1** First reactive boat in the world invented by Alexandra Churcu and Just Buisson, in two variations, on the Seine River in 1886 (engraving in "la Nature", 1887).

In the autumn of 1886, the Head of the French Military Arsenal, Mr. Maurouard* participated as a specialist-guest in one experiment carried out by Churcu and Buisson with their reactive boat on the Sena river; in his report prepared for the War Ministry, the French expert appreciated the basic principle of the device and the original solution adopted by the inventors.

> "... You can imagine a big rocket horizontally installed on the rear part of a vehicle i.e. a small boat or a balloon-cabin, in such a manner that the slow combustion of a special mixture produced exhaust products, going free at the back of the mobile. Moreover, you can suppose the rocket closed in a cannon; in this case, the burning gases exhaust only by firing through the cannon's mouth, producing the well-known reactive motion of the cannon.
>
> If for example, the cannon is fixed on a boat, this reactive force would be transmitted to it and would produce its movement using only the reaction and no propeller or sails.
>
> The inventors installed on their boat a cylindrical container in which a special mixture discovered by them was burned in a closed interior and produced a quantity of gases without any solid residue.
>
> In the rear part of the cylindrical container was a hole destined to exhaust the reactive gases; the hole section was variable, using a manual actuating valve. The pressure, controlled by that valve in the cylindrical receiver, had its values displayed on a manometer. The reactive force manifested by exhausting gases caused the boat to advance continuously-- about 15 minutes--in the opposite direction of the waters of Sena river ..."

Considering Mr. Maurouard's competence and probity, his report was the best qualified proof of the successful test with the first reactive boat.

We can add there were two test versions of this pioneering jet engine: the first with a single cylindrical container, and the second, more perfected and amended by its inventors, with two containers; one for burning the propellant and another for exhausting gases.

## PATENT NO. 179001/12.10.1886

The report of the French War Ministry expert Mr. Maurouard concluded with the possibility that the Ministry buy the Churcu and Buisson invention; the two inventors amended the 'jet engine' and prepared their own patent, issued by the French Ministry of Commerce and Industry with the number 179001 on 12 October 1886.

Unlike the major part of the reactive vehicles conceived in that period, which didn't go beyond the level of a paper draft, *Churcu's and Buisson's jet engine was built and tested in various versions*: One was installed on a small boat and tested as the first reactive boat in the world (3rd of August 1886); another powered the first in the world reactive rail car, which was run in 1887 near Paris at Sevran Livry. *The third type was converted into a special jet engine for aeronautical purposes, being supplied with a reactive jet-oriented system, designated to ensure perfect controlled flight and vertical take-off*. Though the inventors' main purpose was 'air conquest', lack of financial support stopped any possibility of testing this last version.

---

\*   One of Mr. Maurouard's descendants, Mr. Guy Mitaux-Maurouard is a test-pilot on Mirage 2000 N aircraft for Dassault-Breguet.

The original form of Churcu's and Buisson's French patent (text & drawings) wasn't known in Romania until 1982; the only bibliographic source used was Churcu's scientific article published in the renowned French journal *La Nature* (No. 735 dated 2nd of July 1887).

With the assistance of Mr. Guy Buisson and Mrs. Churcu-Stroja (died in 1983), the youngest of Churcu's daughters, the Romanian scientist-writer I. M. Stefan obtained for us a copy of this very important document; it contains 10 pages of text and 3 drawings (in the original patent there were four drawings). The two Romanian specialists Stefan and Zăgănescu were in touch concerning all the details on Churcu's life and scientific achievements dissemination.

In Patent No. 179001/86 are some important scientific formulae for that period, i.e.: Very clear and correct technical expressions for: a) the principle of reactive motion; b) the reactive force formula; c) the independence of the reactive force value of the environment, as follows:

> "The jet engine invented by us uses reaction produced by the forcible gases exhausted through a small hole made in the container; this reactive force propels the container in a direction opposite to that of the gases projection. This propulsion is proportional to the force of gases and to the hole section; it is independent of the environment where the gases exhaust."

The combustible used was given the importance deserved:

> "The source of the gases was the combustion of a special propellant in a closed container; this mixture for which we claim to be our exclusive own is composed of 78% ammonium nitrate and 22% petroleum (kerosene); after these two components were closely mixed, we added 7% wood-carbon, well fragmented in an ammonium nitrate solution . . ."

> *"Because our propellant is a mixture of some substances which are able to supply each other all the elements necessary for their combustion in a closed container*, after the charge was lighted, it will be fully converted into gases, in the absence of every contact with the environment."

The possibilities open to such a jet engine to be built in various versions depending on the destination, all being based on the same principle, were described:

> "We have represented, as an example on the attached drawings, a version of our invention, but this device in which we burn our combustibles, and the gases that are developed, may have various shapes, dimensions and disposals for its main parts, all depending on its destination."

The technical solution chosen by the owners of the patent demonstrated that their invention had all the main components of a conventional rocket engine, such as combustion chamber, nozzle, propellant supplier, burners, after-burner system, lateral jet ducts etc.

> ". . . In two cylindrical containers, made of plate steel, took place alternately the close combustion of our special propellant. One of the two ends of those containers is mobile to assure the propellant supply. The tightness of these mobile doors was ensured by screws and a metallic set. All container-burners (similar to the combustion chambers of the rocket motor, n.a.), communicate with the third container, made also of plate steel, having the role of a gas generator; the volume of this third container is equal to the volume of the other two containers. Communication between the three containers is open only in the combustion period.'

"The gases produced in the first two containers should be in large quantity in all three closed chambers immediately after the propellant was lighted."

"In less than one minute, the pressure of gases becomes 15 bars; at this moment the valve communicating with the reactive pipes open to obtain the reactive force."

Special attention was given to the chemical reactions of the combustion process, especially so that no residue in the container-burners, should be deposited:

"The contents of gases are ammonia, carbon-dioxide, nitrogen, hydrocarbides and no solid residue. In the presence of hot metals, ammonia decomposes and doubles its volume, attending the following chemical reaction:

$$4NH_3 = 2N + 6H^*$$

We use this characteristic to amplify the power of our propellant."

## CONCLUSIONS AND ACKNOWLEDGEMENTS

The last test before the contract was to be signed with the Ministry of Civil Navigation, was prepared for the 16th of December 1886. Unfortunately, when the reactive boat (supplied with a new set of valves and other unchecked systems), was on the Seine river near Clichy bridge, one container of the jet engine exploded: Just Buisson and a young technician were fatally wounded, and Churcu was wounded. The tests undertaken in 1887 at Sevran Livry with a rail car on which the tireless inventor Churcu mounted a new version of his jet engine were successfully run and this was the first jet-car in the world, before the rocket-cars built and tested by Max Valier and Fritz von Opel.[†]

In 1889 Churcu was responsible with all the preparations for the Romanian Pavilion at the Universal Show in Paris. The success was complete and Churcu received permission to return to his country.

One of the original versions of the jet engine was stored at the inventor's house in Bucharest, 13 Labirint Street, and after the death of the inventor, 22nd of January 1922, at the Technical Museum in Bucharest.

## ACKNOWLEDGEMENTS

The authors would like to express their gratitude to Mrs. Rodica Stroja-Churcu and to Just Buisson's nephew, Mr. Guy Buisson, who made available the main data of this paper. The authors acknowledge also the support of the National Center of the Romanian Aircraft Industry and also the Academy of the Socialist Republic of Romania, both in Bucharest.

---

\* In the original form of the patent, this chemical reaction was written as follows: $4AzH^3 = 2Az + 6H$.

† In Otto Willy Gail's work entitled Mit Rakettenkraft ins Weltenall, mention was made that the first car with rockets was built in 1928 by Fritz won Opel.

# REFERENCES

1. Brevet d'Invention No. 179001 du 12 Octobre 1886, pour un Propulseur à Réaction, delivré par le Ministère du Commerce et de l'Industrie de la Republique Française (Patent No. 179001/12.1o.86 for a Jet Propulsion Engine, issued by the Ministry of Commerce and Industry, Republic of France).

2. Ciurcu, Al., La possibilité de la propulsion par réaction (Jet Propulsion Availability). In: *La Nature* No. 755, 02.07.1887.

3. Din viața lui Alexandru Ciurcu (On Churcu's life). In: The Newspaper *Dimineața* No. 3559/30 ianuarie 1914.

4. Vailiu-Belmont, G., Propulsia prin reacțiune. In: *România Aeriană* an.V, nr.44 pp.17-19, iunie 1931 (Jet Propulsion: Aeronautical Romania, June 1931).

5. Moroianu, D., Stefan, I. M., Focul viu. Pagini din istoria invențiilor și descoperirilor românești. București, Ed. științifică, 1963 (The Living Fire, Pages on the Romanian inventions and discoveries History; Scient. Publishing House, Bucharest, Romania).

6. Zăgănescu, F., Della Icar la cuceritorii Lunii. București, Ed. Albatros, 1975.

7. Zăgănescu, F., Gheorghiu, C., Historical Data on the Romanian Industry, AVIATION, Technical Publishing House, Bucharest, 1981.

8. Zăgănescu, F., Ailes Roumaines (Romanian Wings), In: Romanian Review, XXXV year, No. 6-7, 1981.

9. Stefan, I. M., Cel dintîi brevet de notorietate mondială. In: *Contemporanul* nr. 21/21.05 și nr. 22/225 1981 (The first universal valuable Romanian Patient, In: newspapers *Contemporanul* - Bucharest, 1981).

10. Zăgănescu, F., Alexandru Ciurcu. In vol. *Personalități românești ale științelor naturii și tehnicii*. Ed. științifică și enciclopedică, pg. 107, București 1982. (Alexandru Churcu. In the volume: Romanian personalities of the natural and technical sciences. Scientif. and encyclopedic Publishing House, Bucharest, 1982).

# Part IV

# ROCKETRY AND ASTRONAUTICS AFTER 1945

AAS 91-288

Chapter 8

## COMMUNICATION SATELLITES: THE EXPERIMENTAL YEARS[*]

### Burton I. Edelson[†]

"It may seem premature, if not ludicrous, to talk about the commercial possibilities of satellites," wrote Arthur Clarke in 1957. "Yet the aeroplane became of commercial importance within 30 years of its birth, and there are good reasons for thinking that this time scale may be shortened in the case of the satellite, because of its immense value in the field of communications."[1]

Good reasons indeed! Only a few weeks later the Soviet Union launched *Sputnik 1*, and the United States soon followed with *Explorer 1* and a number of other scientific and applications spacecraft, many involving communications experiments. Technical and economic feasibility were soon demonstrated. As a result, not thirty, but only eight years elapsed before the first commercial satellite, "Early Bird", entered service; and in just twelve years commercial satellite service extended around the globe and became profitable.

How was it that the communications satellite gained commercial value in such a short time? Three ingredients were necessary for this to happen, and fortunately all three were, or shortly became, available: technology to create the system; communications requirements to form a market; and a management structure to implement the system. This paper treats the development of all the technologies through experimental and developmental satellites that made satellite communications not only possible, but eminently practical and profitable, in a very short span of time.

Many different technologies are needed to create a communications satellite system. These flowed from diverse sources. Launch vehicle and spacecraft technologies came from work supported over many years by the U.S. Department of Defense (DoD) and for a shorter time, but more intensely, by the National Aeronautics and Space Administration (NASA). The communications and electronics technologies, on the other hand, came mostly from civil and commercial sources, as did the system design and engineering. The first communication satellites were launched in the United States on NASA Delta vehicles which were

---

[*] Presented at the Seventeenth History Symposium of the International Academy of Astronautics, Budapest, Hungary, 1983.

[†] Associate Administrator for Space Science and Applications National Aeronautics and Space Administration, Washington, D.C. 20546, USA, Fellow of AIAA.

derived from earlier military rockets. The spacecraft technologies for the first commercial satellites came from NASA experimental spacecraft; and the communications technologies were inherited largely from commercial microwave systems. The first Earth stations found their roots in radiotelescopes, radars, and microwave transmissions systems. The technologies of satellite communications are truly eclectic and their origins and development paths interesting indeed.

## FORMING THE CONCEPT

It is Arthur C. Clarke, the internationally known science fiction writer, then a Royal Air Force officer, to whom we are principally indebted for the concept of using Earth orbiting satellites for telecommunications. In 1945, he recognized the unique nature and usefulness of the geostationary orbit[2] and suggested "space stations" in that orbit as a means "to link TV services to many parts of the globe." He foresaw (incorrectly) a rather limited future for terrestrial radio links and cables as means for distributing communications services. "A relay chain several thousand miles long would cost millions," he wrote, "and transoceanic services would still be impossible."

A few years later in the United States, John R. Pierce of Bell Laboratories, writing without knowledge of Clarke's words, independently suggested several promising system configurations employing passive as well as active satellites in low altitude as well as geostationary orbits.[3] He raised a point, which has since proved the very basis for the ascendance of the satellite over the cable for transoceanic telephony: Looking at the first 36-channel submarine cable just then (1955) being laid under the Atlantic by the British Post Office and American Telephone and Telegraph Company and costing some $35 million, Pierce asked, "would a channel 30 times as wide, which would accommodate 1080 phone conversations or one TV signal, be worth 30 x 35 million dollars?" He recognized this then as an absurd thought and predicted (correctly) that a technical solution could be found before the commercial demand reached that point. Satellites, of course, have proved to be the solution and we can now provide a thousand channels and more--not at a cost equivalent to many cables, but for only a fraction of the cost of one cable.

With the concepts of Clarke and Pierce in place, the launch capability demonstrated by several Sputniks and Explorers and electronics technologies becoming available, it was possible in the early 1960s to envision an operating satellite communications system. Still, the means and organization for developing and operating a system were not apparent.

Then, in 1961, President John F. Kennedy took two considered steps to accelerate the pace of the U.S. space program. One, quite famous now, was his announced goal to send an astronaut to the Moon in that decade; the second, not so well known, was his "Policy Statement on Communications Satellites."[4] His statement recognized the potential value of satellites to provide communications services; established government policy to coordinate activities and carry out research and development; called for implementation by the private sector and, most importantly, invoked an international effort with these words: "I invite all nations to participate in the communications satellite system, in the interest of world peace and closer brotherhood among peoples of the world."

President Kennedy's bold and prescient statement led to passage by the U.S. Congress of the Communications Satellite Act of 1962[5], which in turn, set in motion a series of events which resulted in the formation of the Communications Satellite Corporation (Comsat) in 1963, and INTELSAT in 1964, and the initiation of the global system in 1965.

It is interesting to realize now that the 1961 statement was based largely on prediction and promise, very little on the results of development or on demonstrated technology. At the time, no active satellite had flown to test voice or video transmission, no spacecraft had ever been put in geostationary orbit, no data on reliable operation of electronic devices or rotating mechanisms in space were available, and some considerable doubt existed as to the acceptability of long-delay transmission paths for commercial service. Still, engineers and managers pushed onward, and as each technical, economic, operational, political, and organizational problem unfolded, they seem somehow to have found solutions.

As the stage was being set for the initiation of commercial satellite communications, a number of particularly knotty technical and operational questions posed themselves. Those engineers and planners in DoD and NASA who were involved in developing technology and launching experimental satellites and those members of Comsat's technical staff responsible for designing and specifying the initial system, faced and tried hard to solve these key problems:

o Should the satellites be active or passive?

Previous experimental work using both the Earth's natural satellite, the Moon, and the metallized plastic balloon, Echo, had shown that signals could be bounced successfully off a passive reflector in orbit and received on Earth. But a very high level of transmitted power would be required and the aperture of the receiving station would have to be quite large. An active electronic receiver, amplifier, and transmitter on the satellite would provide much more system gain. But could this electronic instrumentation withstand the trauma of a rocket launch and survive the orbit for the months or years necessary to make the system economically viable?

o What orbit should the satellites be in?

With relatively low-powered rockets, rudimentary guidance systems, and limited orbital control capabilities, the geostationary orbit suggested by Clarke, though desirable, looked very difficult to attain in the early 1960s. Test flights had been at altitudes of a few hundred kilometers. Then too, the question was raised (largely by experienced system engineers at AT&T) as to whether the long delay time involved in transmission through a satellite in geostationary orbit (over 1/2 second for the two-way trip) would be tolerated in commercial telephone service. There was concern both for the inherent delay and for the ability to suppress "echo" on the circuit. However, since a geostationary satellite system could provide global coverage with three satellites, whereas a medium altitude (say 3000 to 5000 kilometers) would require 20 or more satellites; and since the former would require just one Earth antenna at each location, while the latter would need two steerable antennas--there were powerful technical and economic reasons for going to a geostationary satellite system if it proved technically possible.

o What frequency band should be used?

Various experimental military and civil satellites had used frequencies in the VHF and UHF bands for tracking, telemetry, and command functions; but for telecommunications it would be more efficient to use higher frequencies in the "microwave" region (above one gigahertz) where a great deal more bandwidth would be available, and therefore, a higher information transmission capability attained. Microwave transmitting and receiving equipment was then available from radar systems for ground use, but not for space. Could it be made available? Particularly, could light, efficient, and reliable power output tubes be developed for spacecraft use? If so, which frequencies in the one to ten gigahertz band should be used for transmission up? and down? Because of the power output tube challenge, it seemed desirable to use the lower frequency on the down link.

o What services should be provided?

The obvious advantages of satellite communications over other forms lie in their wide area coverage, broad bandwidth, and direct access to small terminals. These advantages would seem to make satellite communications particularly useful for television broadcast and mobile services. Indeed, television broadcast was originally suggested by Clarke. However, the commercial infrastructure already existed for transoceanic point-to-point telephone service, and a market existed in the early 1960s, at least in the Atlantic region.

## SATELLITE EXPERIMENTS

The years from 1958 to 1964 were the true "experimental years" for satellite communications. During this crucial period, technology developed rapidly. At the beginning of the period, little technology existed, no service had been tested in orbit, and not one of the questions raised above could be answered. But by the end, enough technology was available to provide confidence to answer those questions and to create an operational satellite communications system which became INTELSAT.

### Passive Satellites

Even before Sputnik, design and experimental work had been conducted on passive satellite communications. The U.S. Navy began testing Moon relay communications in 1954 and started an operational service between Washington, D.C., and Hawaii in 1959.

A metallized balloon satellite was first suggested for tracking purposes at the London Congress of the International Astronautical Federation in 1951. In 1956, William J. O'Sullivan at the Langley Research Laboratory of the National Advisory Committee for Aeronautics (NACA--later to become NASA) proposed to develop a 12-foot diameter version of such a balloon. His concept came to the attention of John Pierce of Bell Laboratories and William H. Pickering, Director of the Jet Propulsion Laboratory of CalTech, who suggested it as a passive communications satellite. First, the DoD turned down the opportunity to sponsor this project preferring to concentrate on active satellites which seemed more likely to meet their requirements. But then, NASA, in one of its earliest decisions, decided to support the concept, which resulted in the Echo program.

Leonard Jaffe, an engineer, who had just come from the Lewis Research Center in Cleveland to be Chief of Communication Satellite Programs at NASA Headquarters, saw the potential value in both passive and active communication satellites. He was responsible for directing the Echo program and all the active satellite development work that NASA was later to pursue and which developed the many technologies that eventually became the foundation for INTELSAT and many other systems. Jaffe's book, *Communications in Space,* published in 1966[6], is the best source of detailed information on the early development of communication satellites.

Passive satellites were touted as "simple, reliable, and long-lived." The Echo spacecraft were the simplest type of isotropic reflectors. There was no expectation that they would ever develop into operational systems, but their purpose was to test propagation through the atmosphere and ionosphere and to develop transmission techniques. Not so incidentally, they turned out to be valuable in the development of Earth station equipment and technology.

*Echo I* was a 30-meter diameter, metal, and plastic balloon launched in August 1960, into an inclined 1600-kilometer orbit from Cape Canaveral on a Thor-Delta vehicle. *Echo I* provided the first live, two-way voice communications via satellite. Many other optical and radio tests were made using stations around the world. Within a week of its launching, the first transoceanic satellite signal transmitted from Bell Laboratories in New Jersey was received in Paris by an Earth station of the Centre National d'Etudes de Telecommunication.

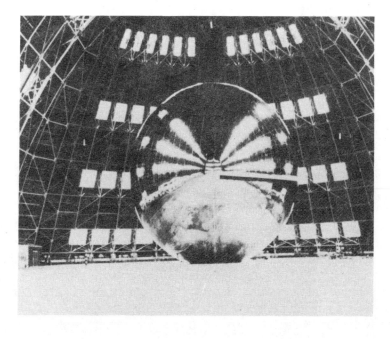

**Figure 1** The Echo-II passive reflector satellite.

*Echo II*, somewhat larger and more rigid, was launched in January 1964, by a Thor-Agena. Tests with *Echo II* included transmission between a station at the University of Manchester, England, and the State University at Gorky, U.S.S.R. The Echo satellites proved the feasibility of radio transmission via satellite, measured propagation characteristics, and demonstrated the effectiveness of various items of transmitting, receiving, coding, and modulation equipment. Perhaps because they were so easily visible to observers on Earth, the Echo satellites excited the general public and stimulated enormous interest not only in satellite communications, but in all the practical applications that space systems had to offer.

Whereas the Echo balloons were discrete reflectors, Project *West Ford*, sponsored by the U.S. Air Force, used an orbital "belt" of small wire filaments as dispersed reflectors. This project was carried out by the Lincoln Laboratories under the direction of Thomas F. Rogers and Walter E. Morrow, Jr. The advantages of a dispersed reflector system over a discrete system were continuous availability and large reflection area, whereas the disadvantages included frequency limitations and interference with other systems. Successful tests of *West Ford* were conducted in 1963. Although no attempt was made to implement an operational system, the project contributed in important ways to developing ground terminal equipment and transmission techniques. Also, it brought the efforts and talents of people like Rogers and Morrow and others at Lincoln Labs working on military research to bear on the field of satellite communications with many future beneficial results to civil systems.

**ACTIVE SYSTEMS**

The first active communication satellite was *Score*, a 60 kilogram payload built in just a few months by DoD and carried into orbit on the side of an Atlas rocket in December 1958. As a "delayed repeater," it received a message from Earth, stored it on tape, and later in another part of its orbit transmitted the message to ground. *Score* transmitted President Eisenhower's Christmas message to the world and became known as the first "voice from space." Its 8-watt VHF transmitter lasted less than two weeks on battery power and died on New Year's Eve.

The second DoD effort was *Courier*, like *Score* a delayed repeater, but this time a separate spacecraft intended to develop an operational capability. *Courier* was spherical, 130 centimeters in diameter, with a mass of 230 kilograms. It was a highly complex, solar cell powered satellite with four receivers, four transmitters, and five tape recorders, processing and repeating several kilobits per second of digital data. It was launched on a Thor-Able-Star vehicle into a 1000 kilometer orbit in October 1960. *Score* and *Courier*, together, proved that delicate and complex electronic equipment could be made to survive the trauma of rocket launch and function in orbit. *Courier* demonstrated all essential subsystems of an active satellite--communications, power, telemetry, and command. It also demonstrated the unfortunate and inevitable tradeoff between complexity and reliability. *Courier* operated perfectly for 18 days, then failed, due to a command system fault.

Encouraged with the success of *Courier* and *Score*, the DoD in late 1960 embarked on a most ambitious project, *Advent*--aimed at placing a 1000 kilogram body-stabilized, high-powered microwave satellite in geostationary orbit. Two years were scheduled for this task. The Air Force assumed responsibility for building the spacecraft; the Army and Navy, for the ground and shipboard terminals. *Advent* was terminated in 1962 after the expenditure of $170 million. The reason was simple: Requirements were set well beyond technological capability to meet them--a lesson to be learned over and over again in aerospace systems development.

Telstar was the most famous experimental communication satellite of all--technical contributions were so significant and its impact on the public so great-that its name for a while became generic for "communication satellite." Telstar was developed by Bell Laboratories at AT&T under the guidance of John Pierce, and with AT&T corporate funding, became the first significant undertaking in space by private enterprise.

**Figure 2**  The Telstar satellite, developed and built by the American Telephone and Telegraphy Company.

Telstar was an 88-centimeter diameter, with a faceted sphere, mass of 80 kilograms, covered with solar cells with an antenna belt around its waist. Significantly, Telstar received at six GHz and transmitted at four GHz bands that later were assigned to commercial service and used by INTELSAT and other "fixed service" systems exclusively during the 1960s and 1970s. Telstar was the first satellite to use a traveling wave tube (TWT) power output amplifier. Its output was three watts. A complete description of Telstar can be found in the *Bell System Technical Journal.*[7]

*Telstar I* was launched on July 10, 1962, on a Thor-Delta vehicle into an elliptical inclined orbit of perigee 950 kilometers, apogee 5,600 kilometers. Its period was 158 minutes. For months, many types of communications tests were conducted between Earth stations especially built or modified for the project at Andover, U.S.A.; Goonhilly Downs, England; and Pleumeur-Bodou, France. These Earth stations were coordinated by the NASA ground station committee formed and chaired by Len Jaffe.

**Figure 3**  The original Earth station at Andover, Maine, used for experimental service with projects Telstar and Relay and later with the operational INTELSAT system.

The day *Telstar I* was launched, live television was transmitted from the U.S. and received in England and France and transmitted from France to the U.S. This spectacular demonstration was followed by many other communications tests over its four months of operation. *Telstar I* demonstrated the specific communications techniques and equipment that would be used in commercial systems. It also showed that a medium altitude satellite system could be used effectively with antenna tracking the many satellites, and communicating, simultaneously.

*Telstar II* was launched in May 1963 into a considerably higher orbit, 10,800 kilometers, giving its period of 225 minutes with relatively long periods of visibility across the Atlantic. *Telstar II* served long and well enough to prove to all that a medium-altitude commercial system was feasible, and to many that such a system was preferable.

NASA's medium-altitude communication satellite, termed Relay, followed Telstar by several months. Although similar in concept, Relay, had two transponders and used ten-watt TWT's. *Relay I* was launched in December 1962, and *Relay II* in January 1964, both into inclined orbits with apogees of about 7,400 kilometers. Both performed very well over several years demonstrating live television transmission around the world, including "firsts," to Germany, Italy, Brazil, Japan, and the U.S.S.R.[8] Several Earth stations, later to become part of the INTELSAT system, first operated with the Relay satellite including those at Fucino, Italy, and Raisting, Germany.

Far and away the most important experimental communication satellite was Syncom--the first to be placed in geostationary orbit. Its design became the model for INTELSAT's first four generations of satellites and for satellites in many domestic systems. The concept of a spin-stabilized spacecraft for use in geostationary orbit was created by Harold A. Rosen and Donald Williams of Hughes Aircraft Company in 1959-60. After DoD turned the mission down, NASA awarded Hughes a contract in August 1961 to design and build a satellite following their concept. Syncom was planned to fly on a Thor-Delta rocket which could put about 70 kilograms into an elliptical transfer orbit with an apogee of 36,000 kilometers. The spacecraft was to be designed with its own apogee kick engine capable of circularizing the orbit and placing about half the weight, or 35 kilograms, into final orbit. The Syncom satellite was packed full of stabilization and control equipment; its communications package had rather limited capacity in two transponders, each with two-watt TWT output stages. Because the Syncom system inherited ground and ship terminals from the defunct Advent program, it had to use their frequencies in the 7/2 GHz bands.[9]

*Syncom I*, launched in February 1963, failed to survive the apogee motor firing; but *Syncom II* in July 1963, and *Syncom III* in August 1964, succeeded admirably and demonstrated the great utility of the geostationary orbit for almost all satellite communications services.

Actually, due to limited propulsion capability, the orbit of *Syncom II* was not in the equatorial plane, but was inclined $33°$ so that it traced a figure eight pattern in the sky. Still, it clearly demonstrated the great advantage to ground stations of minimal tracking and continuous coverage. Communications links once established could be maintained for days or months without interruption. Its obvious use for full-time telephone and live television were quickly accepted. *Syncom III*, using a thrust-augmented version of Thor-Delta, a true geostationary satellite, clinched the argument for most system engineers.

Tests and demonstrations with the Syncom satellites went a long way toward gaining acceptance of the use of geostationary orbit satellites for telephone service. NASA, Stanford Research, AT&T, and the FCC all participated in evaluating the tolerance of users for the half-second round-trip delay along the 150,000 kilometer path from ground station to satellite to ground station, and back. Electronic echo suppressors were developed to limit the amount of original signal of the return path. The result of these tests was that, while critical engineers could not agree, the general public found the circuits quite acceptable.

**Figure 4** Syncom, which became the world's first geostationary satellite.

The Syncom satellites operated for several years performing many experiments and demonstrations, including the first satellite communications to the African continent, to a ship at sea, and to a jet airliner in flight. Syncom started the trend to small antennas, with the Army using a transportable five-meter diameter terminal for two-way voice communications. Although not designed for TV transmission, Syncom brought the 1964 Olympic games from Tokyo to the U.S. After experimental tests, the Syncoms were used by NASA for operational support of the Gemini and Apollo manned spaceflight programs and by the DoD for operational support of the military services.

Syncom set the stage for INTELSAT, the first operational system. It proved out the rocket, spacecraft, and communications technologies needed for geostationary satellites, and it introduced a superb design and manufacturing team in Hughes Aircraft Company to the communications satellite business. Harold Rosen, a co-author of the concept, carried on to become the world's leading designer of communication satellites for over two decades and a principal technical contributor to many systems.

It should be noted that the network of ground stations developed for use with the NASA experimental satellites--Andover, Maine, in the U.S.; Goonhilly, England; Pleumeur - Bodou, France; and Raisting, West Germany - provided the early base for commercial service in the INTELSAT system.

## **Development Satellites**

The INTELSAT system, as the first operational system, benefited greatly from the technologies developed during the experimental years from *Score* to *Syncom*, 1958-1964. However, it was still not clear in the years 1963 and 1964--when Comsat was forming and making initial technical decisions--and even into the late 1960s, what the ultimate INTELSAT system configuration should be.

Some questions were settled early. Active satellites were deemed more effective for commercial service than passive ones. The 6/4 GHz band, tested by Telstar and allocated by the ITU, would be employed. But the orbit question was still open. Although Syncom had proved that the technology would work in geostationary orbit, the question of whether the geostationary orbit would be acceptable for commercial telephone service was not settled. Also, concern was expressed as to whether the electronics, stabilization, orientation and electrical power systems would be suitable for long life in orbit.

INTELSAT considered both medium and geostationary orbit satellites for its initial system. The Comsat technical staff, working then as "Manager" for INTELSAT, observed carefully and critically as NASA and DoD continued their experimental work to test and demonstrate new technologies, to improve spacecraft performance and reliability, and to develop more efficient communications techniques.

In doing this, Comsat was fortunate to have on its early technical staff a number of engineers who had previous firsthand experience with satellite experiments. To mention only a very few: Siegfried H. Reiger, a brilliant systems analyst from the Rand Corporation, headed their technical team; Sidney Metzger from RCA who had participated in projects *Score* and *Relay*, became Comsat's chief engineer and made major contributions to Earth station technology, and Martin J. Votaw, who had built experimental satellites at NRL, became project manager for satellite development. Wilbur L. Pritchard from Aerospace Corporation, with extensive experience developing military systems, was appointed the first director of Comsat

Laboratories and led it to become a principal research center for satellite communications technology. This team created the INTELSAT system using at first inherited technologies from experimental projects, and, later, technologies developed at Comsat labs as part of INTELSAT's own R&D program.

During its first decade of operational service, INTELSAT continued to benefit greatly from the flow of spacecraft and communications technologies from experimental satellites. Many of these came from the NASA Applications Technology Satellite (ATS) program. Six satellites in this series, each designed to test different new technologies, were built and launched by NASA in the ATS program in the years 1967 to 1974.[10] The technologies included electronically and mechanically despun antennas (*ATS-1* and *-3*, respectively), hydrazine propulsion for stationkeeping (*ATS-3*), and the use of large antenna (*ATS-6*). All six demonstrated the use of wide-band multiple-access microwave communications; *ATS-5* introduced K-band, and *ATS-6*, L-band communications. It is interesting that *ATS-2* and *-4*, both intended to be body-stabilized spacecraft, failed to attain proper orbit, and thus, were prevented from demonstrating the advantage of this technique. *ATS-1* and *-3*, both spin-stabilized spacecraft, performed very successfully and are still operating. No doubt this comparison of stabilization systems influenced the future course of satellite design. However, *ATS-6*, a body-stabilized spacecraft, did achieve orbit in 1974 and with its multifrequency 9.1-meter diameter antenna, very successfully demonstrated the use of satellite broadcasting, networking, and data collection with small antennas. The NASA ATS series, all of which were built by Hughes Aircraft Company, were extremely useful in developing and stockpiling technologies that were later to be used not only by INTELSAT, but by domestic, maritime, broadcast, and military systems as well.

The Lincoln Laboratories of MIT was very active in developing technology specifically for the military services. Their work on the Lincoln Experimental Satellites (LES) series was done at X-band and at VHF, but involved advances important to commercial communications, such as transmission efficiency, spacecraft power, high-gain antennas, stabilization and stationkeeping, and the use of small ground terminals.

The LES developments led to the Tacsat program which was designed to provide both experimental tests and operational communication services for the U.S. military forces. An important technology advance in Tacsat was the "gyrostat" stabilization concept which allowed a cylindrical spacecraft to be spun about its long central axis. This concept found immediate application in the *INTELSAT IV* spacecraft which served successfully throughout the decade of the 1970s and into the 1980s.

The *Symphonie* satellite, sponsored by France and Germany, also contributed to communications technology development and demonstration during this period. Symphonie-1, a body-stabilized spacecraft launched in 1974, was very helpful in demonstrating the advantages of satellite communications for regional and domestic services and helped greatly to spread knowledge and capability in this burgeoning field to Europe and around the world.

One other development satellite program made important technological contributions during this period. This was the Canadian Communication Technology Satellite (CTS). Launched in 1976, it used an advanced body-stabilized spacecraft design with lighweight, foldout solar arrays. It carried a high-power, 200-watt, 12-GHz traveling wave tube power amplifier and demonstrated broadcast and "thin route" communication services to very small ground terminals.

## CONCLUSION

INTELSAT went into operation in 1965 having benefited from a very exciting and productive period of development and test of experimental communications satellites from 1958-64. During those six years, all of the spacecraft subsystems--structure, stabilization, control, propulsion, and electrical power; and all the communication subsystems--antennas, wideband receivers, filters, and power amplifiers; were developed, tested, and demonstrated to the extent that INTELSAT could proceed directly into operational service. Then, during its first decade of operations, INTELSAT continued to receive the technological output of many development satellite projects. These technologies carried INTELSAT through four generations of satellites. Then, starting with *INTELSAT-V*, more dependence was placed upon technologies resulting from its own R&D efforts than from outside sources.

INTELSAT was the first and is still the largest communications satellite system. However, the same technologies that allowed international communications traffic to flow across oceans and continents were applicable to other services as well. The Soviet Union was the first to develop a domestic satellite communication system for its own needs. Canada, the United States, Indonesia, and other countries followed. In 1976, a maritime satellite communication system was initiated. Today more than a dozen separate international, regional, domestic, and mobile systems are operational, providing needed, and unusually profitable telephone, television, and data services via satellite.

The whole history of communication via satellite has been a short one. Technologies and systems have evolved in rapid, almost explosive, growth from experimentation, through development and operational service, to a routine business--all in a quarter century. The rapid past progress, the present prosperity, and the glorious future of satellite communication represent a triumphant demonstration of the benefits of technological development in general; and, in particular, the world's investment in space exploration.

# REFERENCES

1. Arthur C. Clarke, *The Making of a Moon*, 1957.
2. Arthur C. Clarke, "V-2 for Ionosphere Research" and Extraterrestrial Relays," *Wireless World*, Feb. 1945.
3. J. R. Pierce, *The Beginnings of Satellite Communications*, San Francisco Press, 1968.
4. J. F. Kennedy, Presidential "Policy Statement on Communications Satellites," July 24, 1961.
5. "Communications Satellite Act of 1962," U.S. Congress, Public Law 87-624, August 31, 1962.
6. L. Jaffe, *Communications in Space*, Holt, Reinhart and Winston (1966).
7. "The Telstar Experiment," *Bell System Technical Journal*, Vol. 42, No. 4, July 1963.
8. R. H. Pickard, "Relay I - A Communication Satellite," *Aeronautics and Aerospace Engineering*, September 1963.
9. Syncom Projects Office, Goddard Space Flight Center, "Syncom Engineering Report," Volume I, NASA TR R-233, Washington, D.C., March 1966.
10. "Applications Technology Satellite and Communications Technology Satellite User Experiments for 1967-1978," prepared for NASA Lewis Research Center by University of Dayton Research Institute, June 1979.

**Note**: This paper is abstracted from the forthcoming AIAA Progress Series Volume "The INTELSAT Global Satellite System," to be published in 1983.

AAS 91-289

Chapter 9

## PROJECT ROVER:
## THE UNITED STATES NUCLEAR ROCKET PROGRAM[*]

### Dr. James A. Dewar[†]

From 1900 onwards, space pioneers speculated that atomic energy could provide an inexhaustible source of energy which would make the exploration of space a reality. In the mid-1950s the United States initiated a nuclear rocket program called Project Rover which would last until 1972, cost over $1.5 billion, and have several potential missions. However, while the nuclear rocket had great potential, it never had a fully approved or defined mission and this in the final analysis caused its termination. This study analyzes the Rover program from technical, managerial, and political perspectives, examines how successes or failures in one of these areas affected the other, and evaluates whether the program was beneficial.

The idea to use atomic energy to propel a rocket for interplanetary space travel existed for over 40 years before the United States made a decision to develop a nuclear rocket. To be sure, this idea was most fanciful during the period 1900-1945 because the sciences of the atom and rocket were in their infancy. After the atomic bombs were dropped on Nagasaki and Hiroshima in 1945, some detailed engineering studies of a nuclear rocket were made. But by the end of the 1940s, there still was no serious interest in any country in nuclear rockets.

### EARLY THOUGHTS ON ATOMIC ROCKETS

Starting around the turn of the century, a number of rocket and space pioneers speculated that atomic energy was the ultimate rocket fuel as it was an inexhaustible and would open up the door to interplanetary space travel. Robert Goddard, the American space pioneer, was perhaps the first to begin this speculation. As a college sophomore in 1906-07, he wrote a paper on the utilization of atomic energy. A gram of radium had the potential to lift 5,000 tons over 100 yards in height, Goddard noted, but its disintegration was so slow that years would pass before enough energy was released naturally to lift even a gram. Atomic disintegration had to be controlled and not occur spontaneously; but once achieved, the

---

[*] Presented at the Seventeenth History Symposium of the International Academy of Astronautics, Budapest, Hungary, 1983.

[†] U.S. Department of Energy, Washington, D.C.

navigation of interplanetary space could begin. Goddard wrote on the subject in 1916, but afterwards concentrated on developing chemical rockets.

During the 1920s and 1930s, a number of space and rocket pioneers speculated on the use of atomic energy. They included Gaetano A. Crocco of Italy, K. E. Tsiolkovsky of Russia, Eugen Sänger, Hermann Oberth and Krafft A. Ehricke of Germany and P. E. Cleator of the United States. However, all of them essentially concluded that atomic energy was a rocket fuel of the future.

The development of the atomic bomb and the German V-2 rocket during World War II prepared the way for serious post-war thought of marrying the two sciences to produce a rocket capable of navigating through space. The first was by Leslie R. Shepherd and A. V. Cleaver of the United Kingdom who collaborated in 1947-48 to produce a remarkable series of papers, using only unclassified information, which were published in the *Journal of the British Interplanetary Society*. They surveyed the various nuclear propulsion systems, including the merits of solid and gaseous core reactors and the use of hydrogen as the working fluid. They concluded that future interplanetary space flight would require nuclear or ion propulsion, but at the moment, nuclear rockets were not technically feasible.

The second two-phase study was prepared by North American Aviation Company in 1946-47. The initial study investigated the technical aspects of using several structural materials and propellant working fluids. For example, with data obtained from preliminary calculations of liquid hydrogen, methane, and water, fairly accurate measurements of the size and weight of a graphite reactor were made and a conceptual design for a 10,000 mile intercontinental ballistic missile was produced. This led to an in-depth six-month study which looked at all conceivable difficulties attending an atomic-powered rocket with a range of 10,000 miles and a payload of 8,000 pounds. The study concluded that liquid hydrogen (LH) was the best working fluid because of its high specific impulse and that graphite was the best structural material for the reactor. However, on the basis of experiments, North American noted that hydrogen eroded graphite at a rapid rate--perhaps an appropriate analogy is a cube of sugar dissolving in a cup of coffee. Thus, North American postulated that feasibility of the nuclear rocket only could be established fully when a method of protecting graphite from hydrogen erosion had been developed. Upon completion of this study, however, further North American interest in nuclear rockets ceased.

## ORIGINS OF PROJECT ROVER

In the early 1950s, the missile and nuclear weapon rivalry grew between the U.S. and Soviet Union. Both sides realized the potential intercontinental ballistic missiles (ICBM) would have for their nuclear arsenals, but each side pursued different courses to realize that potential. The Soviet Union, having exploded its first atomic bomb in 1949 and working on the more powerful thermonuclear bomb, focused its missile effort on developing large rockets capable of carrying its large and heavy nuclear weapons. The Soviet Union did not focus its immediate attention on making its nuclear weapons smaller and lighter. On the other hand, the U.S.

focused its atomic bomb development (and thermonuclear weapons development program) on making them smaller and lighter. Thus, a missile would not have to have a large boosting capability in order to deliver its payload. However, the principal U.S. ICBM in the early 1950s, the Atlas, encountered a number of technical difficulties and was characterized as a plumber's nightmare. At the same time, development work had not conclusively proved that U.S. nuclear weapons could be made smaller and lighter. It was in this atmosphere of technical uncertainty that serious interest in a nuclear rocket developed in U.S. government and national laboratory circles.

## TECHNICAL ASPECTS OF THE NUCLEAR ROCKET

This interest was caused in large measure by an article which appeared in the December 1953 issue of the classified *Journal of Reactor Science and Technology*. Robert Bussard, a young scientist working in the Oak Ridge National Laboratory, reinvestigated some of the earlier studies, such as those by Shepherd-Cleaver and by North American Aviation. He concluded that a solid core, heat-exchanger nuclear rocket engine using hydrogen, methane, or ammonia as the working fluid would be superior to chemical systems for all but the smallest payloads (less than 1000 pounds) and the shortest ranges (less than 1000 miles). The margin of superiority of a nuclear rocket over its chemical counterparts became greater as the payloads became heavier and the distances longer.

This article generated considerable interest in government circles in Washington and in the nation's atomic laboratories. At the Atomic Energy Commission's (AEC) Los Alamos Scientific Laboratory in Los Alamos, New Mexico, several study groups started independent of each other to investigate the concept while at the Lawrence Livermore Laboratory in Livermore, California, a group called the Rover boys was formed to start work. During 1954, studies continued at both laboratories, but as they progressed the advantages of a single stage nuclear rocket over its chemical counterparts were not found to be as great as originally thought. Thus, prospects for the commitment of large amounts of money for development would not be forthcoming. Faced by this prospect, Los Alamos hit upon the idea of boosting a nuclear rocket with a chemical rocket lower stage. After theoretical calculations were made, the results were most important: a reasonably sized rocket but with a very large payload advantage. Clearly impressed with this military potential, the Air Force felt some justification existed for continuing nuclear rocket work. And throughout 1955, the Air Force worked within the Department of Defense (DoD) to formally establish a program which the two weapons laboratories had already informally established. In November 1955, the DoD sent a letter to the AEC requesting it to pursue further research into nuclear rockets.

During 1956, both Los Alamos and Livermore pursued their work aggressively as they were in competition as well, as they were most interested in this new concept. They conducted experiments on candidate reactor core materials and working fluids, developed designs for various nuclear rocket engines and airframes, and surveyed and developed designs for a test site in a remote desert in the state of

Nevada. In Washington, however, it was becoming apparent to decision-makers that a two-laboratory nuclear rocket program would be a very costly endeavor. As a consequence, the DoD undertook a high-level review of the nuclear rocket program to determine what mission it could fulfill. This review concluded that as the plumbing problems of the Atlas and other missiles were being solved and as there had been great progress in making nuclear weapons smaller and lighter, there was no plausible need for a nuclear rocket for ICBM applications. However, this review also concluded that a nuclear rocket had great potential for future space applications and that the program should be redirected and be pursued at a moderate level of effort toward the goal of demonstrating the feasibility of the concept. From a political perspective this change was most important. In Washington politics where there is continual fighting over money, it would be very easy to criticize any money spent on a nuclear rocket development program. However, it is not quite as easy to criticize a program which has as its goal the demonstration of feasibility of the concept.

This guidance was provided by the DoD to the AEC in January 1957. The AEC then reevaluated the programs of the two laboratories in the light of this guidance. Henceforth, all nuclear rocket work would be conducted at Los Alamos. Livermore was assigned the task of working on Project Pluto, a nuclear-powered ramjet. Ironically, although Livermore's nuclear rocket work was reassigned, their division nickname, Rover, became the code word for the project at Los Alamos.

To implement this new guidance, the group at Los Alamos, under the direction of Raemer Schreiber and Rod Spense, decided upon a basic reactor testing effort called the KIWI program. Named after the flightless New Zealand bird, KIWI was a two-phase effort, moving from the relatively easy to the more difficult steps in establishing feasibility. KIWI-A, the easier, was to determine the basic data both in reactor physics and materials and in testing procedures. KIWI-B, the more difficult, was to use liquid hydrogen (LH) as the working fluid. Progressing thus, the many unknowns of reactors reaching temperatures of over $2000°C$ in a matter of seconds could be learned in a methodical, logical, but aggressive manner. Feasibility could be demonstrated with great assurance then.

KIWI-A had a design power level of 100 megawatts (MW). (One MW would provide 1000 pounds or 4.45 KN of thrust). It also had a reactor core 4 feet in diameter and 4 feet in length. The fuel elements were constructed out of flat plates and were called whims, a contraction of the words wheel and rim. These whims were curved and stacked upon each other in the cylindrical core. The working fluid was gaseous hydrogen. While design and fabrication of KIWI-A was underway, other Los Alamos personnel were at a site in the Nevada desert supervising the construction of the elaborate facilities needed to test the KIWI series of reactors. The test site as well as KIWI-A were completed and checked out in early 1959. In July, the first high-power, hot test occurred (Figure 1). Despite the fact that there was a failure of a key part in the reactor core which allowed hydrogen to escape without entering the core, KIWI-A was successfully tested at 70MW for over 5 minutes before being shut down. This test was highly important because it not only technically demonstrated the feasibility of the concept, but it also, as shall be made evident subsequently, was interpreted politically as evidence of a new and most promising energy source for rockets that needed to be developed immediately. There were several tests of other reactors in the KIWI-A series in 1960. KIWI-A and KIWI-A3 were tested to gain further experience in field testing procedures or investigate the performance of prototype fuel elements (Figure 2). However, the KIWI-A series were never viewed as prototypes upon which to build a design for a nuclear rocket engine.

**Figure 1** KIWI-A in full-power operation, on July 1, 1959. Photo from movie frame taken 500 yds away.

After KIWI-A, Los Alamos began serious work on the KIWI-B series of test reactors. These would be far more difficult and challenging than the KIWI-A series. They would have a 1000MW design power level, a drastically different reactor core design based ultimately on hexagonal fuel elements made of graphite-uranium mixture, and would use liquid hydrogen not only as the working fluid but also as the coolant for the rocket nozzle and reactor core. Furthermore, the KIWI-B series had more stringent target performance specifications, 1000 MW of power for 5 minutes.

**Figure 2** KIWI A Prime was operated at full power on 8 July 1960.

Using liquid hydrogen presented a vastly more complex series of technical problems than KIWI-A. For example, Los Alamos determined that over 50,000 gallons (227 meter$^3$) of liquid hydrogen would have to be pumped into the reactor in order to meet the target specification. They built two 28,000-gallon (127-meter$^3$) dewars along with the appropriate plumbing system at the test site to supply this working fluid to the reactor. This had never been done before. Next, the nozzle as well as the reactor core would be cooled with liquid hydrogen. This presented severe thermal problems. In addition to this, there were major worries whether 'slugs' of liquid hydrogen would enter the reactor core; as hydrogen is an excellent neutron moderator, there were concerns that these 'slugs' could cause serious reactor control problems. In other words, the power level of the reactor could not be maintained under control. Finally, as liquid hydrogen would enter the nozzle and reactor core at -253°C and, in the span of 5 feet, leave at a temperature of over 2000°C, there were severe thermal stress, thermal expansion, and structural integrity problems. To assist in solving some of these problems, several private industrial firms were brought into the program; foremost of these at this time was Rocketdyne which developed the nozzle and liquid-hydrogen pump.

It took about two years to redesign the test site to handle liquid hydrogen and about the same time to develop the new KIWI-B series of reactors. KIWI-B1B was the first reactor run on liquid hydrogen (in September 1962). It demonstrated that there were no problems with 'slugs' of liquid hydrogen entering the core. However, the B1B suffered structural problems and ejected a number of fuel elements from the core. This was not viewed as serious, as the design was held to be deficient structurally even before the test. Rather, it was thought more important to have a test on liquid hydrogen and learn about its handling properties.

The KIWI-B4A was viewed as the core design with great promise; it was thought that the B4A could go through a 5-minute full-power 1000MW test run without suffering any structural problems. The B4A was tested in November 1962 (Figure 3). Automatic programming brought the B4A up to low power and then to high power quickly; again the liquid hydrogen startup was successful. But paralleling the rapid increase in power was a rapid increase in the frequency of flashes of light from the nozzle. On reaching 500MW, the flashes were so spectacular and so frequent that the test was terminated and shut-down procedures begun. Quick disassembly confirmed that the flashes of light were reactor parts being ejected from the nozzle. Further disassembly and analysis revealed that over 90% of the reactor parts had been broken, mostly at the core's hot end. The test of KIWI-B4A had not only technical consequences, but also, most important, managerial and political consequences.

**Figure 3** KIWI B4-A was tested successfully on 30 November 1962.

## MANAGEMENT OF PROJECT ROVER

After the program was formally established in 1955, the Air Force was assigned responsibility for the non-nuclear aspect of a nuclear rocket. For example, it would be responsible for taking the reactor engine developed by the AEC and integrating it into the rocket vehicle. However, when NASA was created in 1958, the Air Force responsibilities were transferred to it. As can be expected, there were the normal bureaucratic problems of who would run the program--the developing agency or the using agency. These problems, however, only hid a more fundamental problem in that both agencies had different approaches to research and development. NASA favored a methodical, systematic approach to developing new technology, testing components rigorously before testing the entire system. Because many of NASA's developments would be used for manned airplane or space flight, dependability and reliability were emphasized. In this context, time was sacrificed in order to minimize risk. In contrast, the AEC had an aggressive approach to research and development, originating with the weapons development work in the 1940s and 1950s. Here rapid weapons development was paramount, with cost and risk sacrificed in order to save time. Thus, different approaches to technical problems were initiated and continued in parallel until one proved superior. Component and system testing, in developing a new technology, were conducted especially when there was a good opportunity for failure. Learning the unknown was more important than relearning the known. In this context, reliability and dependability were goals to achieve later in the development process.

In 1958 and 1959, this was not a serious problem, as the nuclear rocket had not been demonstrated to be feasible. However, after KIWI-A which demonstrated feasibility, this problem became more critical, particularly after NASA published several long-range planning documents for its future activities in space. The nuclear rocket was not mentioned, nor was it assigned missions 20 years or more in the future. This was troublesome to the AEC and infuriating to some of the Rover program's most staunch supporters in the Congress. At this time, the Congress of the U.S. was controlled by the Democrats who, after the Soviet Union had launched Sputnik in October 1957, continued to press for a much larger U.S. role in space. Some of the Democrats pressing for this larger role were also members of the Joint Committee on Atomic Energy (JCAE) (composed of members of the House of Representatives and Senate who had the responsibility to oversee the development of atomic energy). The JCAE sought a much more aggressive nuclear rocket program--as fast as the technology would allow. When NASA's long-range plans had no near-term mission for a nuclear rocket, the JCAE realized that it meant that the necessary funds to build test facilities, to bring in private industry, in essence, to develop a nuclear rocket, would not be forthcoming.

Under JCAE pressure the management problems were partially solved in 1960 when a joint AEC/NASA office was formed. It was modeled after the very successful joint office that was created to develop the U.S. nuclear submarines under Admiral Hyman G. Rickover. A controversy developed over who would be named to head this office, a NASA or an AEC man. Ultimately, NASA prevailed and Harold Finger from NASA's Lewis Research Center in Cleveland, Ohio, was named to head the office. Finger was schooled in the NASA development philosophy, but was not able to impose that philosophy upon the AEC until after the KIWI-B4A test in November 1962. Afterwards, Finger decided there would be no further hot tests until the cause of the core failure had been determined precisely and the solution to the problem tested repeatedly under cold test procedures before any hot testing would be resumed. Cold flow testing had a policy effect as it meant a hold on the other portions of the Rover program which were aimed at flight testing a nuclear rocket.

Nonetheless, throughout the first part of 1963, cold testing was done on another KIWI-B4A type reactor. On a heavily instrumented B4A was a specially designed camera which was inserted into the nozzle to take motion pictures of the cure during the test. In startup, as the pictures indicated, the gas flowing though the core caused severe vibrations which cracked the fuel elements. Some were ejected. Convinced, on the basis of the pictures and other data that vibration was the problem, corrective redesign of the KIWI-B4 series began. In August 1963, a redesigned KIWI-B4B was cold-flow tested and the test was completely successful. No fuel elements were cracked or ejected. Thus approval was given for the resumption of hot testing.

Beginning in January 1964, work began toward testing of the last two of the KIWI-B series of reactors, KIWI-B4D and B4E. In May, the B4D was hot-tested. Starting quickly and completely automatically, the B4D reached and maintained full power of 1000MW for about a minute until a leak in the nozzle forced termination of the test. Other than the nozzle failure, disassembly confirmed that the test was a complete success. The core was intact, no vibration had occurred. The core design was good. (Opinions still vary on whether the ban and subsequent cause and effect cold-flow testing procedure was warranted. Some adamantly maintain that a year and a half was lost in proving what Los Alamos had suspected originally as being the design fault. Little therefore was gained except interesting pictures. On the other hand, others staunchly hold that development steps followed prior to the ban on hot testing were too cavalier to produce an engine safe and reliable enough for man-rated flights. Little confidence in the soundness of an engineering design can be gained from failures).

The last of the KIWI series, the B4E had the same core design but featured an improved method of coating the fuel elements. Tested in August 1964, the B4E proved the most successful KIWI. Running for eight minutes at 900MW, the duration of the test was limited by the storage capacity of the liquid hydrogen dewars. Startup and control were smooth and stable. The core performed well with no flashes. The exit gas temperature was 2000°C, slightly lower than B4D. After shutdown, the B4E was not disassembled and analyzed. Rather it was decided that valuable information could be acquired on fuel element lifetimes by going beyond ten minutes in total reactor running time and on reactor restarts and reliability. The B4E was restarted and run at full power for two and one half minutes and then shut down. No problems were encountered. Subsequently disassembly and analysis revealed that the new fuel elements suffered only minimal corrosion and that the core remained intact. The reactor could have run much longer.

## THE POLITICS OF THE ROVER PROGRAM

The Rover program was throughout its life a creature of partisan politics. It was strongly supported by the Joint Committee of Atomic Energy in the Congress, and in particular, by a few powerful Democratic senators who sat on the committee. These senators were also members of close friends of senators who were members of the 'inner sanctum' of the Senate, those senators who really wielded power. Senator Clinton P. Anderson, a Democrat from the state of New Mexico, was the Rover program's strongest supporter and if he was not a member of the 'inner sanctum' he was a close personal friend of the one person who was the head of it as well as the Senate, Senator Lyndon B. Johnson from Texas.

Anderson was unhappy in January 1957 when the DoD stated that were no missions planned for a nuclear rocket, but that it should be pursued at moderate level of effort to demonstrate the feasibility of the concept. However, at that time he had insufficient grounds upon which to press for a larger program. After the Soviet Union launched its Sputnik satellite in October 1957, the political climate in the U.S. dramatically changed. Now the Democrats raised the issue that the U.S. was losing the space race and they pressed for a much larger U.S. response. The Eisenhower Administration, however, did not feel that the space race represented a tangible threat to U.S. security. Nevertheless, it felt that it had to respond to the criticism of the Democrats. Throughout 1958, hearings were held in Congress on the creation of a civilian space agency, NASA, to run the nation's space activities. At the end of the year it was officially created and a number of military programs transferred to it. The Air Force responsibilities in the Rover program, as noted previously, were transferred to NASA. While some of these newly transferred military programs were given increased funding as NASA began to establish its programs and priorities, the Rover program was not given an increase; in fact, it was not even mentioned, or mentioned only 20 years in the future, in the planning documents that NASA was developing at this time. As the KIWI-A test, to demonstrate feasibility, had not occurred, there was little the Democrats in Congress could do to accelerate the program.

After the KIWI-A test in July 1959, the AEC formulated a budget for an aggressive program leading to a flight test in the mid-1960s timeframe. This was submitted to the Bureau of the Budget, the agency which formulated the Administration's budget for transmittal to Congress. The Bureau of the Budget disallowed the AEC's budget request and kept the program funded at a level of effort sufficient only to do further feasibility demonstration work.

When this budget became known in January 1960, when it was transmitted to the Congress for review and approval, it aroused many Democratic members, particularly Senator Anderson. He had expected the project to be accelerated following the KIWI-A test, and he moved to determined action. In February 1960, he notified NASA and the AEC that he had scheduled executive session hearings to cover the following points on the Rover program: to establish firm operational objectives, a flight test schedule, and a management structure suitable to accomplishing those objectives. Immediately upon receiving Anderson's letter, NASA sent a letter to the Bureau of the Budget supporting the AEC in the program. The following day, the Bureau notified NASA and the AEC that the original budget request was partially amended and that the AEC could reprogram from its other funds to make up the remainder, if it wanted. In the closed hearings that news was conveyed to Anderson; however, the hearing produced no new NASA position on flight dates, objectives, or organization structure.

Nonetheless, after receiving Senator Anderson's letter, both agencies began discussions to determine their reaction to the strong Democratic emphasis. But with a Republican Administration conservative toward space, both NASA and AEC leaders were limited in their freedom of action. Thus, they could not establish officially approved flight dates and firm objectives. In view of this situation, the questions discussed in the Rover program in the early half of 1960 were essentially organizational. Centralized management had been discussed in the AEC and NASA since the space agency was created, but as the nuclear rocket was viewed as a long-term development effort in NASA, the need to establish a joint office to manage and coordinate the program was not considered important. The Democratic emphasis changed this attitude though. In early April 1960, NASA and the AEC agreed on forming a Space Reactors Branch, headed by a NASA man, in the AEC's Division of Reactor Development. The proposed arrangement was shown to Senator Anderson, but he questioned the plan, favoring an office modeled after the one headed by Admiral Hyman G. Rickover who ran the U.S. nuclear submarine development program. Here one man with real authority ran the program. That type of office had considerably more power than the one proposed by the agencies. Shortly afterwards, a new plan was proposed, based on the Rickover model, and was accepted by Senator Anderson and the Joint Committee. As indicated, Harold Finger was named to head this office.

Throughout the summer of 1960, the staffs of the two agencies worked to implement the agreement; by late August 1960 a memorandum of understanding was signed between NASA and the AEC establishing a Nuclear Propulsion Office. However, this office did not have anything meaningful to do at this time because a full-scale, flight-test oriented nuclear rocket program had not been approved. That decision was being left to the next Administration.

Having a national decision on a program on the magnitude of the nuclear rocket program was necessary for policy and managerial reasons. On the national level, expending anywhere from one-half to one billion dollars envisioned in 1960 to develop an operational nuclear rocket meant a tacit commitment to use that vehicle in the space program. But constructing a nuclear rocket, capable of moving very large weights in space, implied a commitment to a very expanded space program aimed at unmanned or manned space--even planetary exploration. This fact was appreciated by many Democrats. And at the party's nominating convention of 1960, John F. Kennedy, who was least knowledgeable about space, was nominated for President; Lyndon B. Johnson, who was very well informed about space was selected to be Vice President. Senator Anderson was somewhat disappointed as he had backed Johnson for President. Nonetheless he supported both candidates strongly. Although Presidential nomination politics dominated the convention, key

Democrats inserted a plank in the Democratic platform, calling for the development of the nuclear rocket as part of an accelerated space program.

Kennedy won the election but upon assuming office in January 1961 was not predisposed to a large space program. International events placed the President on the defensive and forced him to change his position. The Soviet success in orbiting a man and recovering him safety on April 12, 1961, the flight of Yuri Gagarin in Vostok, was transformed quickly and effectively into a worldwide political, military, and ideological message. Hailed as a triumph of socialism over capitalism and as an illustration of Soviet military strength used for peaceful purposes, the political meaning of the Gagarin space feat was not lost on the developing nations: Socialism was propagated as the wave of the future. Discouraging as this Soviet feat was and embarrassing as the bungling of the Cuban Bay of Pigs was to the President, Kennedy decided that he had to initiate a positive policy in part to redeem his campaign promise of getting the country moving again. After some initial hesitation, the President decided that for political, military, and ideological reasons, the Soviets had to be challenged and surpassed in space, a substitute program like desalinization of water did not have enough international prestige or visibility. Kennedy then assigned to Vice President Johnson the responsibility for making recommendations concerning the scope and direction of the space program on April 19, 1961.

The following day, Johnson began arranging meetings and hearings in order to determine what the scope of an accelerated program should have and how much political support that program would have in the government and the nation. Since Sputnik, landing a man on the Moon had been considered by many Americans to be the proper goal for a space program. In the last days of April 1961, the manned lunar landing became the favorite objective of many government and industry figures in an accelerated space effort. In discussing an enlarged effort, though, Johnson spoke with many informed people who considered a number of other programs and projects which would enable the U.S. to continue space exploration and not reduce all activity after a lunar landing. Foremost among those projects was the nuclear rocket because it was justified as having planetary or lunar resupply capabilities. In this context, developing a nuclear rocket in the 1960s would extend American leadership in space well into the 1970s and 1980s.

For the next six weeks Johnson worked on developing a space policy for the nation and submitted it to the President in mid-May. Kennedy reviewed and approved without change the recommendations given him by Vice President Johnson. On May 25, 1961, President Kennedy addressed the Congress and asked the nation to commit itself to an all-encompassing space program, having as its central objective the landing of a man on the Moon and returning him safely to Earth before the decade was out, and to developing a nuclear rocket which when completed might take men to Mars, perhaps even to the end of the solar system itself.

The following day, NASA and the AEC translated the President's policy into specifics, that the President had decided favorably on the flight test as an objective for the nuclear rocket and that NASA now was authorized to develop a flight rated nuclear rocket engine and to integrate that engine into a rocket vehicle. The target date for a flight was set at 1966-67. Progress towards realizing the 1966-67 flight date now was only limited by technical factors--how fast and how successfully could Las Alamos test the KIWI-B series of reactors on liquid hydrogen. (At this time, NERVA became approved and would be the logical successor to the KIWI series of reactors. NERVA stood for Nuclear Engine for Rocket Vehicle Application).

Thus, the tests of the KIWI-B series of reactors were more than just technical matters viewed only by scientists and engineers. They became the symbol for a large and expansive space effort after the lunar landing. However, no such missions were approved in the President's speech; they were just alluded to. But the full implications of the President's leadership in space policy started becoming readily apparent in 1962 as the budgets of the AEC and NASA came under increased scrutiny. To flight-test a nuclear rocket by 1966-1967 meant that hundreds of millions of dollars had to be allotted in 1962. This to many critics was only the tip of a very large iceberg because if the nuclear rocket were flight-tested successfully there would be even more pressure to use it for far more exotic space missions, perhaps a manned flight to Mars. For this reason determined efforts were made to reduce Rover's size and scope.

The failure of the KIWI-B1B and B4A tests in September and November 1962 were used most effectively by the critics of an expanded space program. They wanted to delay the funding for the NERVA and for the flight test of a nuclear rocket on the grounds that the technology was not ready. To help offset this criticism, Senator Anderson arranged for President Kennedy to visit Los Alamos and the reactor test site in Nevada before making this decision on the budget. This visit occurred during the first week in December 1962, about a week after the KIWI-B4A test.

In the midst of a technical briefing on Rover, the President interrupted to state he wanted to discuss the serious budgetary problem his administration faced with the very large flight test program being proposed by NASA. Kennedy stated he wished to listen to the arguments for and against supporting the Rover program at the projected level. Harold Finger defended the flight test objectives, noting that while the Saturn-V, being developed for the lunar landing program, was being designed on the basis of chemical propulsion, nuclear rockets had influenced its design. Furthermore, it might prove very important in landing men on the Moon before 1970 should there be a failure in the Saturn-V system. Essentially, however, Finger noted that nuclear rockets combined with Saturn-V's figured prominently in NASA's planning for the missions of the 1970s. The President's Science Advisor, Jerome Wiesner, countered, stating that the failure of the KIWI-B1B and the B4A demonstrated that nuclear rocketry was technically premature, that additional basic reactor research was necessary before starting NERVA and the flight-test programs. Rather, NERVA should be oriented to a low-level technology demonstration effort since NERVA was an expensive and technically formidable effort compared with the KIWI program. The flight test should be canceled. Essentially, Wiesner continued, the nuclear rocket was premature and could not be considered useful even as a backup for the lunar landing mission. Mars and lunar base applications were well beyond the pace for any serious government planning--perhaps a generation away from Presidential approval.

The following morning, the President flew to the Nevada test site to visit the nuclear rocket test facilities. Sitting on his bed on *Air Force I*, the Presidential plane, Kennedy again discussed the nuclear rocket's funding and mission applications. The President's advisors again restated their views, that the nuclear rocket was premature, that unless the lunar base program was approved, where nuclear rockets would make one hundred trips a year to the Moon, or unless a manned Mars mission was authorized, the expense of developing nuclear rockets could not be justified. Just before landing in Nevada, the President decided to delay the flight-test program, pending the outcome of the KIWI-B4 tests.

This issue remained unresolved throughout the nine months of 1963. However, as the budgets of the AEC and NASA were being developed in the Fall of 1963, the objectives for the nuclear rocket and by implication for the post-lunar landing space missions for NASA were being debated. Kennedy was mindful that several of his other key programs were entering the expensive hardware development stage, the Minuteman and Polaris nuclear weapon systems and the lunar landing program. All would require significant increases in funding. Reflecting the buildup, the total budget loomed to just over $100 billion, but the funding level for Rover had not been determined when Kennedy was assassinated in November 1963.

On assuming the Presidency, Lyndon Johnson embarked on a course to renew confidence in the government and in part this desire was reflected in his decision to pare the budget below $100 billion. In this context, Johnson considered not only the Rover program, but also the entire space program. In mid-December, there was a meeting of President Johnson and the heads of NASA and the AEC. They discussed three funding levels for Rover: a $300 million per year level aimed at a fight-test objective; a $200 million per year level aimed at flight-rated engine, but no decision on a flight test; and a $150 million-per-year level aimed at research and technology, with no flight-rated engine development of flight test. Johnson ruled out the $300 million-per-year option. Thus, discussion centered on the $200 million-per-year versus the $150 million-per-year option. There were some Presidential advisors present at this meeting who advocated a $70 million-per-year effort. No decision was reached at this meeting, but it was apparent that the third option had emerged as the leading choice.

The next week was spent in weighing the political implications of the $150 million research and technology plan, essentially with key Democrats in Congress. There was a second meeting during Christmas week, but it was apparent that Johnson had reached his decision before the meeting began. The Rover program would be reoriented to a research and technology effort at a funding level of $150 million per year; NERVA would be redirected to a ground-testing reactor program and the flight test terminated.

The nuclear rocket program went on to be an outstanding technical success. Liquid hydrogen ceased to be a problem and the reactor fuel elements were developed and improved. Reactor control and operation techniques were enhanced by Los Alamos, which also proceeded to develop a 5000MW reactor suitable for planetary missions and by the private industrial firms working on NERVA that developed a series of continually improved designs for their experimental reactors (Figure 4). Repeated starts and stops were practiced as well as running for long periods of time at full power. In 1968, a NERVA reactor ran at full 1100MW in power for one hour without damage. Experiments to improve the fuel element lifetimes to five-ten hours also were started; an engine with that capability would be a very cost-effective vehicle for a lunar base. But such missions were never approved. In 1972, with Senator Anderson in ill health and in his final year in the Senate, the Rover program was terminated by a Republican Administration. There simply was no need for a nuclear rocket as there were no missions for it.

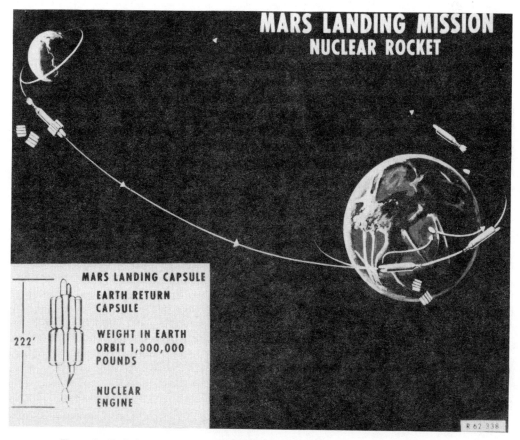

**Figure 4** Typical 1962 planetary mission concept where reactor power of first stage leaving Earth orbit is about 10,000 Megawatts.

## AN EVALUATION OF PROJECT ROVER

The Rover program was an unqualified technical and managerial success. Technically, very severe engineering problems with a graphite-liquid hydrogen heat-exchanger nuclear rocket were tackled and solved in a relatively short period of time--about 15 years. And in the final years of the program there was successful work on developing a nuclear rocket capable of multiple restarts. Had the program not been terminated in 1972, there is no doubt that there would have been further technical improvements. Perhaps there might be even better second-generation nuclear rockets. However, the task of developing a flight-rated system would have remained even if the Rover program were not terminated. This would have posed a number of challenging engineering problems, but not insurmountable ones.

From a managerial perspective, there were some early differences of views with respect to the AEC's and NASA's approaches to research and development. However, after the KIWI-B4A test, these problems settled down. During the last ten years of the program, the joint AEC/NASA office was firmly in charge and successfully managed the efforts of the AEC and NASA laboratories and the private industrial firms who were brought into the program. Everyone worked as a team to produce the technical successes which were obtained in the mid and late 1960s.

However, the Rover program faced insurmountable political problems. It was the favorite program of an influential but aging Democratic senator and when he left office, there was no one in the Congress who could effectively muster support for the program. Had there been a mission for the nuclear rocket though, that support might have been found in the Administration or the Congress. But it was most difficult to justify funding a program which was aimed at developing a rocket engine for advanced space missions--a lunar base or a manned Mars expedition--when NASA's space activities after the lunar landings were being severely curtailed.

If U.S. attitudes change, as they might, for there are indications of renewed public support for manned space exploration, the nuclear rocket program could be reestablished. If it is, there will be a solid body of technical successes and managerial know-how to build upon, perhaps leading to improved performance second-generation nuclear rockets. Then President Kennedy's statement might be fulfilled, that the U.S. would have a nuclear rocket to "take men to Mars, perhaps to the end of the solar system itself."

AAS 91-290

Chapter 10

# A COMPARATIVE STUDY OF THE EVOLUTION OF MANNED AND UNMANNED SPACEFLIGHT OPERATIONS[*]

## Kristan Lattu[†]

Twenty-six years ago, spaceflight operations consisted of little more than engineers and scientists detonating projectiles that were virtually uncontrolled after launch. More "operations" dealt with range safety, test activities, and tracking the projectile. Today, there are well-coordinated tracking networks and control centers around the world which conduct spaceflight operations, control and attendant data processing. The evolution of space operations and ground support has, however, been relatively unknown. Spacecraft hardware and mission results tended to receive the largest share of attention. Information detailing certain aspects of mission operations has been preserved somewhat inconsistently. Following the end of a project or mission, documentation has often been lost unless it related directly to the mission results or problems under investigation. Test data, operations staffing and team budgets have frequently been buried in files, though total NASA manpower and budget figures are widely publicized. "What" was accomplished has been carefully archived, while "how" it was accomplished and supported--a whole unique field of aerospace--has been virtually ignored in the various space histories. The disciplines of space operations and control now involve several specializations, innovations in hardware and software, and an interesting variety of similarities and differences between the manned and unmanned programs.

While several space operations facilities have evolved in the U.S. and elsewhere, I will focus on the manned space operations center in Houston and the unmanned operations center at the Jet Propulsion Laboratory. The discussion concludes with some speculations on future space operations.

### BACKGROUND AND COMPARISON

What is meant by "spaceflight operations and control"? In the broadest sense, we would say that operations consist of all activities which influence and control the spacecraft and the mission. In fact, operations have been considered to be those

---

[*] Presented at the Seventeenth History Symposium of the International Academy of Astronautics, Budapest, Hungary, 1983.

[†] Jet Propulsion Laboratory, California Institute of Technology, 4800 Oak Grove Drive, Pasadena, California 91109.

activities which were done on the ground (because in the early space program all spacecraft control originated on the ground). In particular, a few functional areas seem to combine with a homogeneity and developmental history that lend themselves to treatment as a field. These functional areas are as follows:

- A. Operations Planning - the design and integration of ground and spacecraft events, simulation, resource scheduling, testing and training

- B. Tracking and Data Acquisition/Relay - tracking spacecraft, relaying command and/or telemetry data between ground control centers and spacecraft

- C. Command and Control - including flight dynamics and navigation, monitoring, and other realtime support activities.

The origins of spaceflight operations and control can be traced to the early rocket experiments of the 1940s and 1950s. Scientists and engineers tried to find effective equipment and methods for tracking and controlling their rockets. Then, as now, ground operational developments were driven by a need to effectively support a technical experiment. The primary objective in innovation and change of ground systems has been to obtain and process more data more quickly and more reliably, both to and from the spacecraft, in order to control the spacecraft more effectively.

Manned space operations were at first centralized in military flight test groups working on experimental aircraft at Langley Research Center. These groups were the source of the original astronauts, and of ground monitoring and support personnel as well. A control center was established at Langley and at the launch complex at Cape Canaveral, but in 1962, the post-launch control center was moved from Langley to Houston and called the Manned Spacecraft Center (now the Lyndon B. Johnson Space Center - JSC)[1]. Another center of operations was developed at the Goddard Spaceflight Control Facility (GSFC) which became the lead center for the Space Tracking and Data Acquisition Network (STADAN) and for control of Earth orbital missions. It also served as a routing facility for communications lines to overseas tracking facilities for all three tracking networks. GSFC now manages the STADAN and the Manned Spaceflight Network (MSFN) and the combination is called the Space Tracking and Data Network (STDN). It provides tracking and data relay resources for both manned and unmanned Earth orbital missions. GSFC provides communications links to the Deep Space Network (DSN) for its domestic and overseas facilities through the NASA Communications Network (NASCOM) circuits.

The functions and facilities of Jet Propulsion Laboratory (JPL) were transferred to NASA in 1958. The DSN, managed by JPL, was officially established in December, 1963, with four 26m antenna stations and a Spaceflight Operations Facility (SFOF - located at JPL), which utilized the first block of IBM 7094 computers.[2] An important difference between the manned space operations and JPL's unmanned operations was the early arrangement of JSC as a separate organizational entity with a single objective of providing ground operational support to manned spaceflights. Because JPL also built some elements of the unmanned spacecraft and its components, and was a research center for many of the experimenters, the conduct of operations was inextricably mingled with functions and personnel not truly trained for operations. The JPL operational environment was also made more complicated in the early mission by a variety of administrative difficulties between NASA, JPL's parent institution (California Institute of Technology), and contractors and scientists from other institutions.[3] Each project felt the unique characteristics of their project required tailored operations. It was thought that standardization would reduce flexibility and quality for the unique mission. Scientists and experimenters did not always realize when requesting that commands be sent to their instruments what effect those commands might have on other experiments or the overall condition of the spacecraft. At JSC, operations personnel also had to accommodate science interests, but it was always clear that the welfare of the humans, "getting a man to the Moon and safely returning him" were the primary objectives. The focused goal helped to simplify some operational interfaces.

## OPERATIONS PLANNING

With the first long space mission, the unmanned Mariner Venus 1962 (MV62), it became apparent that planning for operations was just as important as designing the spacecraft hardware or the science objectives. The lunar Pioneer and Ranger 1 through 4 missions prior to MV62 had only required about three days of operations, with ground activities during the flight that could be noted on a few sheets of paper.[4] The three-month mission of MV62, however, required more comprehensive planning. A full time communications supervisor was appointed with the task of planning the MV62 and subsequent Ranger communications operations; this operational technique helped provide the capability to support simultaneously the MV62 mission and the Ranger 5 mission in October, 1962.[5] This also represents the beginning of what can be called the field of spaceflight operations: Integrating complex activities involving several systems or ground support with sufficient mission duration to require multiple shift support.

Operations for unmanned Earth orbital spacecraft developed rapidly during the decade following MV62. Planning and scheduling of the required tracking and data relay facilities and the attendant communications and computer resources, quickly became a major operations planning task for unmanned space projects. As unmanned satellites proliferated, both in Earth orbit and deep space, competition for the tracking and data relay resources forced projects to designate full-time staff members to design and negotiate the tracking requirements. For example, during 1973, the mission workload for the combined tracking and data relay networks consisted of 43 unmanned spacecraft, including three international satellites. In addition, the stations provided tracking support for four manned missions, and for instrumentation left on the Moon by previous Apollo missions.[6] In a typical week of 1973, the workload for the Merritt Island Station (used for pre-launch test activities, launch tracking and on-orbit support) served 81 spacecraft/mission activities.[7]

The manned program generally had less competition for ground resources, since manned missions tended to be sequential rather than simultaneous. However, launch and landing activity planning were challenging tasks because flight launch and reentry have to compete with on-going Earth activities such as civilian air and shipping traffic. Mission event planning was complicated by having to integrate engineering and science activities with biological and medical activities associated with humans on-board. The training of the astronauts became a major and integral function of the manned space operations planning process. In comparison, the unmanned missions put top priority on exhaustive system tests, first with the separate flight instruments, then integrated tests of the whole spacecraft. This was necessary because once launched, the unmanned probes would have no humans nearby to effect repairs.

Another comparative element in operations was that the majority of the unmanned spacecraft could not be recovered and analyzed. Therefore, "System Test" was important to the unmanned mission as a means of not only testing the spacecraft systems to ensure they were functioning correctly, but also to train the ground engineers in how the spacecraft would respond, and its various idiosyncrasies. Then, when the spacecraft was in flight, remote operations personnel would understand how to operate more efficiently, or "fly" the spacecraft. The manned program emphasized analysis of the post-flight space capsule and astronaut observations as a means of improving the subsequent flights.

It is interesting to note that around the time of the first extended manned missions (*Gemini 4*), the manned space operations groups first began to function in 24-hour shifts. Operations teams were trained and used during several missions of similar characteristics. Personnel learned to treat the various processes in a more routine and standardized way. The characteristics of orbital flights and the repeatable conditions contributed to the opportunity to refine operations techniques. The focused objectives of the manned missions also contributed to the development of more standard operations, compared to the often competing interests of an unmanned mission, where multiple experiments might be under the authority of various institutional organizations. The more complex and diffused objectives in unmanned deep space missions resulted in each mission and its operations being unique.[8] The tracking and data relay systems, however, which were managed separately as a multi-mission service, were able to move toward standardization more easily.

Perhaps as a by-product of the more cohesive structure of manned operations, personnel in that operations arena tended to remain from mission to mission, building up the overall expertise of the manned operations organization, even with normal attrition. A different situation developed in the unmanned program. With the advent of the very long deep-space mission, such as *Pioneers 10* and *11*, and the Voyagers, a special attrition problem emerged, associated with the extended duration times. For example, some Voyager personnel began working in the initial planning phases in 1972; by the time of launch, five years later, many had gone on to other positions. When *Voyager 2* reaches its final solar system encounter with the planet Neptune, in 1989, some 17 years will have passed since the first design work. When *Voyager 2* departs the solar system, hardly any of the original Voyager operations personnel will still be working on the project. Most of the people who participated in Voyager's pre-launch systems tests, and gained valuable experience operating the spacecraft, will no longer be with the project.

## TRACKING AND DATA ACQUISITION/RELAY

The origins of tracking and data acquisition go back to Germany where Doppler tracking had been used to track the early V2 rockets. With the development of early missiles and rockets came the evolution of radiometric techniques for tracking and control.[9] JPL designed and built a tracking network called Microlock, which consisted of Doppler-telemetry mobile stations, while under contract to the U.S. Army. This system was used to track *Explorer 1*, even though the Navy had a somewhat similar system called Minitrack. The crucial difference was a JPL innovation called the phase-lock loop receiver which provided a more accurate signal track and improved radiometric knowledge. Later, this feature was included in all tracking station equipment.[10] The first Pioneer lunar probes required a different data acquisition capability to receive the weaker signals of a small spacecraft transmitter at a great distance. So larger, steerable antennas were built at fixed base sites, and, within a few years, the DSN, STADAN, and the MSFN were established under the Office of Tracking and Data Acquisition (OTDA).

The evolution of the tracking and data relay systems and their managing organizations proceeded along different directions from those of the control centers. From the start, these networks were organized and managed separately from the flight control groups. The objectives of the OTDA organization were focused and oriented to serving multi-mission uses. Technological advances, especially in the reduction of signal noise, precise timing standards, and computerized tracking, helped stations to perform more efficiently and to to develop more standardized procedures and automated systems. In 1972, when the MSFN combined with the STADAN to form the Spaceflight Tracking and Data Network (STDN), the first step toward integrating manned and unmanned operations was accomplished. Later, automation also contributed to overall efficiency in this area. Because of the speed of orbiting spacecraft and the Earth's rotation, the tracking of past missions was limited to constrained view periods. At first, reconfiguration of the station in order to acquire different spacecraft was performed manually. This took time and limited the tracking time still further. As the stations have become more automated, the time of "turn-around" between spacecraft has been reduced.

Although the DSN and JPL began working to mold unmanned space operations as early as 1965 into a more efficient, "multi-mission" oriented service,[11] each new unmanned deep-space mission brought pressures to create project-unique command systems or revised telemetry modes and data processing requirements. Handling of Earth orbital unmanned satellite traffic, however, did become more routine and heavily automated during the last decade as orbital conditions and operational requirements became fairly well-known and predictable.

The key element tying the tracking function into the operations area has been that of communication between the spacecraft and the control centers. For the unmanned program's very distant probes, a major improvement has been to "array" stations. This technique improves the data reception of distant, weak signals. The combination of arraying, together with image compression coding, should enable *Voyager 2* at Uranus (over a billion miles from Earth) to have a picture return rate of almost 30 Kbps (at Saturn it was 44.8 Kbps), whereas it was once expected that no more than 20 Kbps could be achieved.[12] Stations arrays may be further utilized in the future to enable smaller, less expensive stations to achieve a reception of weak, distant signals comparable to those of large-dish antennas.

The other major development in this operations area is the advent of the Tracking and Data Relay Satellite System (TDRSS). This set of satellites will give an almost continuous link from ground stations to on-orbit activities in the Space Shuttle. The unmanned spacecraft *Galileo*, destined for launch in 1986, may use the TDRSS following deployment from the Space Shuttle until it is can be acquired by the DSN. This system promises to add versatility to spacecraft ground communications with continuous coverage capability of Earth orbital spacecraft.

## COMMAND AND CONTROL

Paralleling the development of the tracking systems was the growth of the communications and computing centers which interfaced with them. These centers became the focal point for control and commanding of the spacecraft. Some of the greatest differences between manned and unmanned space programs can be seen in the area of command and control.

In manned operations, the critical control link was the voice communication between the ground and the astronaut. Some control over the manned spacecraft was exercised by the on-board human, while other commands were telemetered from the ground. Unmanned spacecraft, until recently, had to be entirely commanded and controlled from the ground. Table 1 compares the various ground and on-board functions of past, present and future manned and unmanned operations.

In general, deep-space unmanned missions require far more intensive command sequence planning. Very long one-way transmit times for command (due to speed of light at interplanetary distances) make real-time response difficult or even impossible. The manpower and associated costs for ground command planning and integration were powerful incentives (as planetary budgets dwindled) toward the development of spacecraft computers capable of being pre-programmed with an array of responses, selectable by the on-board command program according to various conditions which might fit its parameters. This type of spacecraft can protect itself to a certain extent (from Sun damage, for example) and can be programmed to respond to a variety of contingencies until the ground controllers are able to communicate directly with the spacecraft. In the manned program, control is also shifting from the ground into the spacecraft. However, instead of contingency sequences and autonomous capability, the objective of the manned spacecraft computers is to allow more control, even programming capability by the on-board astronauts. The shift in the relationship between ground and spacecraft control is shown in Figure 1. At some point in future manned missions, it is not unreasonable to expect constant access between ground-based data systems and in-flight computers for navigation, experiment management, and activity planning. When manned missions venture beyond the region of Earth and Moon, different techniques will need to be developed.

Table 1
COMPARISON OF GROUND AND SPACECRAFT FUNCTIONS

| TASKS | MANNED | | | | | | UNMANNED | | | | | |
|---|---|---|---|---|---|---|---|---|---|---|---|---|
| | GROUND | | | ON-BOARD | | | GROUND | | | ON-BOARD | | |
| | PAST | CURRENT | FUTURE | PAST | CURRENT | FUTURE | PAST | CURRENT | FUTURE | PAST | CURRENT | FUTURE |
| NAVIGATION ANALYSIS | TOTAL | SHARED | BACK-UP ONLY | NONE | SHARED | TOTAL | TOTAL | SHARED | SHARED | NONE | SHARED | SHARED |
| SPACECRAFT SYSTEMS ANALYSIS | TOTAL | SHARED | SHARED | NONE | SHARED | SHARED | TOTAL | TOTAL | TOTAL | NONE | SOME PRE-PROGRAMED | SOME PRE-PROGRAMED |
| GUIDANCE & FLIGHT CONTROL | TOTAL | SHARED | BACK-UP ONLY | NONE | SHARED | TOTAL | TOTAL | SHARED | SHARED | NONE | SHARED | SHARED |
| ACTIVITY PLANNING | TOTAL | SHARED | SHARED | NONE | SHARED | SHARED | TOTAL | TOTAL | TOTAL | NONE | NONE | NONE |
| ANOMALY RESPONSE | TOTAL | SHARED | SHARED | NONE | SHARED | SHARED | TOTAL | SHARED | SHARED | NONE | SOME PRE-PROGRAMED | SOME PRE-PROGRAMED |
| GROUND RESOURCE SCHEDULING | TOTAL | TOTAL | TOTAL | NONE | NONE | NONE | TOTAL | TOTAL | TOTAL | NONE | NONE | NONE |
| TRACKING | TOTAL | TOTAL | TOTAL | NONE | CARGO/TARGET TRACKING | CARGO/TARGET TRACKING | TOTAL | TOTAL | TOTAL | NONE | PROGRAMED TARGET TRACKING | PROGRAMED TARGET TRACKING |

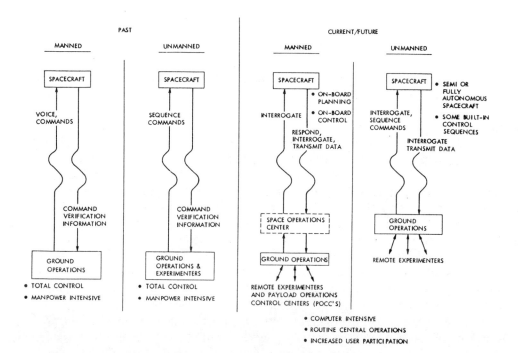

Figure 1  Comparison of Spacecraft to Ground Relationship.

To compensate for the slower data rates being received at very great interplanetary distances, the unmanned space program has evolved advanced data-compression coding techniques for the spacecraft.

In the manned program, emphasis has also been placed on more rapid data return. Earth orbital data rates have gradually increased from 51.2 Kbps, used during the Apollo program, to the projected megabits per second from the Spacelab vehicle to be carried aboard the Space Shuttle. Processing and disseminating this data has become another major task at both manned and unmanned control centers, since user satisfaction with the data results is of paramount importance. Industry advances in computer technology have provided more and faster processing capabilities even as image data and other data types have become more complex. In addition, experimenters now may have their own processing equipment at their remote locations, so that data may be relayed directly from the spacecraft to the experimenters, allowing near-real-time data analysis to take place.

We are, in fact, at a turning point in the history of command and control of space missions. In the past, nearly every activity in either manned or unmanned spacecraft had to be decided upon and then commanded from the ground. With VLSI technology bringing more powerful compact computer systems, we have reached the era of "smart" or autonomous spacecraft. In addition, plans are being made for fully automated tracking and data relay on the ground, and for increasingly automated command and control centers. Further, the growth of commercial and private satellite industries may see command and control shifting to private/remote center, with perhaps only routine tracking and data relay performed by the current control and operations centers.

## CONCLUDING COMPARISONS AND FUTURE OPERATIONS

One of the most significant comparisons between manned and unmanned space operations is the trend in staffing. During the embryonic period of manned space operations, the fairly short (1 day) Mercury flights were controlled with a minimum number of people, all of whom were effectively co-located. By the time of the Gemini series (1964-1966), the simple procedures and interfaces of Mercury had developed into a hierarchical system using one main control area with separate support rooms around it. The numbers of people involved in direct Mission Operations Control Room (MOCR) support, had grown from approximately 40-50 in the Mercury program to almost 100 in Gemini. These staffing numbers reflect those involved in direct mission control and support tasks. By the time the initial Apollo flights were in progress in 1966, Mission Control activities involved more than 250 people around the clock. By *Apollo 17*, that number was further increased by separate staffs of control people who monitored various portions of the flight and flight hardware, such as EVA equipment, rovers, and science instrument packages. The peak operations manpower occurred during the Skylab mission, when the count rose to about 350 people. However, a shift had also occurred in terms of the functions of these people in that they were counted in the operations groups, yet were monitoring and analyzing science instruments aboard the Skylab. Finally, in the Shuttle era, the first flight, STS-1, utilized about 140 persons in operations and control, and in the mature Shuttle operational phase to come, less than 90 people were projected for on-orbit operations.[13] I do not include manpower figures for personnel involved in the tracking and data relay function in these staffing comparisons, as the organizational nature of the OTDA systems was multi-mission rather than project peculiar.

The trend toward more compact operations teams is also reflected in unmanned space operations staffing, although the developmental history differed considerably from the manned program. In the Ranger and early Mariner missions, the individuals who built particular subsystems of the spacecraft made up the early flight operations teams, involving 50 to 80 people in direct support activities. The problem with that approach was that the engineers were less familiar with the total operation of the spacecraft than they were with their separate subsystems, which contributed to the operations difficulties experienced in these missions.[14] More emphasis was placed on integrated tests and training in later missions, and the manpower for flight operations functions was increased. Project operations personnel numbered about 150 for the Mariner Mars 1964 (MM64), then declined for the Mariner Mars 1969 (MM69) to around 50, partly as a result of reductions in the unmanned planetary program, but also because personnel who had worked on earlier Mariner spacecraft were more efficient in fewer numbers due to their experience.

The Surveyor missions used close to 150 in operations, but additional manpower had been injected into the mission because of its importance in supporting the later manned Apollo missions. A peak in operations manpower occurred during the Viking missions to Mars. Operations at JPL were originally limited to the two Viking Orbiters, but later Lander operations shifted from NASA's Langley Center to JPL. Nearly 700 people supported flight operations 24 hours a day, 7 days a week, during the peak operational period of 1976.[15] That number had dwindled to less than 100 by 1979, just as the Voyager mission was approaching its peak staffing of about 350 during the Jupiter encounters. The staffing curve continued down to about 300 for the Saturn encounters and is expected to rise from current low "cruise" level staffing to less than 200 people.[16] Figure 2 compares these approximate staffing numbers for the manned and unmanned operations.

There are several factors reflected in this trend toward smaller operations teams. First, there is enough of a background of knowledge and experience in space that fewer well-trained people can now perform tasks formerly requiring more people who were needed to brainstorm many new and unknown problems. Second, the increased capability and reduced costs of computer systems has made it easier to automate many operational functions. Meanwhile, the increase in private spacecraft has shifted the responsibility for operating them more toward the private sector. In the TDRSS era, with improved direct communications between the spacecraft and the users, we should see a trend toward ground operations involving the users directly.

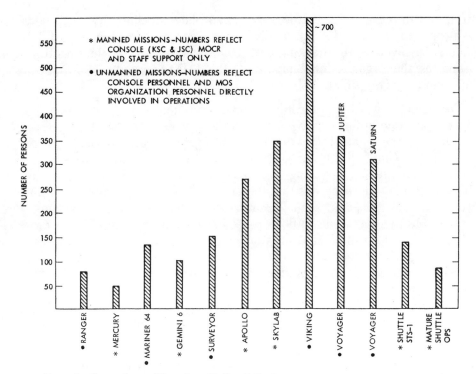

**Figure 2**  Comparisons of Operations Staffing Estimates.

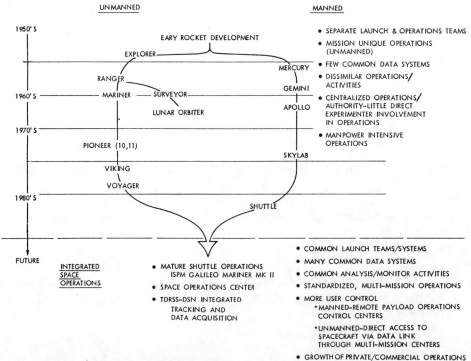

**Figure 3**  Evolution of Spacecraft Operations.

Spaceflight operations show a progressive development from what was a very manpower intensive element of a space mission toward a very computer-intensive functional area. Except for special missions, actual control is becoming distributed to the users. Further, in the mature Space Shuttle operations of the future, both manned and unmanned space vehicles aboard the Shuttle will be data-linked with TDRSS or similar links, and launch and flight schedules will be irretrievably mixed. This growing together of the manned and unmanned programs (as shown in Figure 3) should produce benefits in efficiency and service to the users and researchers in the next phase of space exploration and development.

The next 20 years will bring reduced launch costs, ease of access to low Earth orbit, and possibilities of on-orbit retrieval, repair or even construction. These exciting potentials should stimulate the imaginations of unmanned probe designers and produce yet a new family of robot solar system explorers as well as provide a new workplace in space for people. Spaceflight control and operations may no longer be limited to the ground.

## ACKNOWLEDGEMENTS

The projects and work described in this paper were accomplished under the United States National Aeronautics and Space Administration. I wish to thank the California Institute of Technology's Jet Propulsion Laboratory (JPL), the NASA Headquarters History Office, and the National Air and Space Museum for their assistance in providing access to historical documents and information. I extend special thanks to Frank Hughes of the Johnson Space Center (former Chief of the Flight Training Branch, now heading the development of Space Station training), who was a major source of information and experience with regard to manned space operations, and to James D. Burke, Douglas G. Griffith, and Warren K. Moore, all of JPL, who provided key information and unmanned spacecraft operations."

## REFERENCES

1. Levine, Arnold S., "Managing NASA in the Apollo Era", The NASA History Series, NASA SP-4102, 1982, p.19; and House Subcommittee on Space Science and Applications, of the Committee on Science and Technology, "U.S. Civilian Space Programs 1958-1978", Vol. 1, 97th Cong., 1st Sess., Serial D, Jan. 1981, p.81-83 & pp.92-93.

2. Renzetti, N., "A History of the Deep Space Network - From Inception to Jan. 1, 1969", NASA Technical Report 32-1533, Vol. 1, (Jet Propulsion Laboratory, Pasadena, California, Sept. 1, 1971), p.33.

3. Newell, Homer E., "Beyond The Atmosphere-Early Years of Space Science", The NASA History Series, NASA SP 4211, 1980, Chap.15.

4. Renzetti, p.20.

5. Loc.cit.

6. House Subcommittee on Aeronautics and Space Technology of the Committee on Science and Astronautics, "Review of the Tracking and Data Acquisition Program", 93rd Cong., 1st and 2nd sess., No. 31, (Oct. 24, 1973; Jan. 29, 1974), pp.71, 72, and 83.

7. Ibid., pp.336 and 337.
8. Interview with Dr. Jeremy Jones, July 22, 1983.
9. Corliss, William R., "A History of the Deep Space Network", NASA CR-151915, May 1, 1976, p.2.
10. Ibid., pp.3, 4, 10.
11. Ibid., p.217.
12. Kerridge, Stuart J., "Imaging Science Returns from Uranus and Neptune and Their Sensitivity to DSN Augmentation Planning", Jet Propulsion Laboratory IOM Voyager-SJK-83-1 (internal document), Feb. 4, 1983; and an interview with Douglas Griffith, Voyager Flight Operations Office Manager, Feb., 1983.
13. Johnson Space Center "Flight Control Operations Handbooks": GT-3, Jun. 8, 1964; GT-4, GT-6, Jun. 21, 1965; AS-201, Rev. A, May 24, 1966; Apollo 11, Jun. 6, 1969; Orbital Flight Test Baseline Operations Plan, July 1, 1976; JSC-09-202; STS Baseline Operations Plan, May, 1976; JSC-12267, July. 22, 1983, publ. by NASA JSC Flight Control Division.
14. Rose, Aaron, Ph.D., "Flight Operations Management for Unmanned Planetary Exploration Systems", University of Southern California Directed Research ASM 590, Jan. 1969, pp.77 and 78.
15. Interview with Richard Laeser, Voyager Project Manager, May, 1983.
16. Ibid.

AAS 91-291

Chapter 11

# REACTION MOTORS DIVISION OF THIOKOL CHEMICAL CORPORATION: AN OPERATIONAL HISTORY, 1958-1972 (PART II)[*]

Frederick I. Ordway, III[†]

## TRANSITION FROM RMI TO RMD

Once Reaction Motors had ceased to have an independent existence, the technical alliance with Olin-Mathieson was dissolved, and every effort was made to integrate speedily and effectively the new Division into the Thiokol organization. Dr. Edward H. Seymour, Reaction Motors Division (RMD) manager for much of the 1960 decade, characterized the merger as friendly and "handled as well as mergers can be handled." But, he pointed out, RMD and Thiokol were different kinds of organizations. For one thing, the former was dedicated to the development of liquid propellant rockets and the latter to solid propellant rockets. Also, geographical and cultural differences had to be taken into account. And, of course, there was the matter of size: RMD was a relatively small component competing with other Thiokol elements for management attention and resources.

In Seymour's view, these differences were important and explained why "Thiokol never really fully understood us." The development of solid propellant rockets, he pointed out, is propellant-focused; those involved in their development buy the metal parts (casings) from the hardware store, so to speak, and load them with propellants. In contrast "we liquid propellant advocates dealt with pumping, tubing, ignition, and so on and bought the chemicals we needed from the chemical company." Although solid and liquid rocket technologies have merged somewhat since the late 1950s, important differences in approach persist.

"There was something else," Seymour conjectured, "that would affect a merger anywhere. I'm thinking of the regional background. We were essentially a New York gang. And the Thiokol people were midwestern and southern-oriented. My

---

[*] Presented at the Seventeenth History Symposium of the International Academy of Astronautics, Budapest, Hungary, 1983. **Editor's note**: This and the following article by Frank H. Winter complete this joint history of Reaction Motors from its beginning as an independent company in 1941 through its acquisition by the Thiokol Chemical Corporation in 1958 and eventual demise 14 years later. Part I of the Reaction Motors, Inc. portion of the history appears in Volume 6 (Volume 11, AAS History Series) of the International Academy of Astronautics History Series, edited by Mitchell R. Sharpe.

[†] Space and Rocket Center, Huntsville, Alabama, USA; formerly, Engineering Division, Reaction Motors, Inc.

theory has always been that any organization has an ecology. You force two ecologies together and it is going to take time to merge them. But our working relationships with Thiokol were basically, I feel, very good... We had an extremely capable research organization in physics and chemistry, and to a certain extent, materials. As we saw our market diminishing, I tried to argue that we had capabilities from the extremely basic level on up to applied that was very valuable. Most of Thiokol's other rocket divisions were very good at 'can do' applied work but lacked what we offered. I felt that the company could have capitalized on our experience."

The perspective of Dr. Harold W. Ritchey, at the time of the merger Vice President and Director of Thiokol's Rocket Operations, was somewhat different. When Reaction Motors was brought into its new parent's fold and for the following two years, he told the author, "my objectives were to try to instill a sense of aggressiveness and urgency into the R&D programs of RMD. It seemed to me that ever since Santa Claus came [e.g., Laurence S. Rockefeller, (RMI) benefactor; see Volume 6* in this series] in the early 1950s, the general attitude was that 'It's O.K. - someone will take care of us.'" Such an attitude, Ritchey insisted, would have to be corrected if the Division were to carry its weight and produce a profit for Thiokol. Ritchey had been Technical Manager of Rocket Operations from 1949 to 1957, when he was promoted to Vice President in charge of the same activity. In 1964, he would ascend to the presidency of Thiokol and in 1970 would become Chairman of the Board and Chief Operating Officer.

Seymour takes issue with the thought that RMD lacked aggressiveness and a sense of urgency. "...we certainly went through a lot of organizational changes," he explained, but "... I felt that in general we were a gung-ho outfit and did a good job. Perhaps if I, and others, had had more financial acumen we might have found better directions, but I'm not so sure."

Along with what Ritchey termed "a lot of headaches associated with technical and management problems of the rapidly expanded rocket business," he was faced with line responsibility for the new liquid propellant-oriented RMD. The division, he prognosticated at the time, "appeared to be heading for the biggest headache of all." To many, it appeared to be too narrowly market-based while at the same time suffering the same geographical limitations--a heavily populated East Coast location--that plagued its RMI predecessor. Also, for a variety of reasons, RMD never seemed to "fit" comfortably into the Thiokol corporate umbrella. Its legacy as a small, independent, technically driven research and development enterprise was not easily accommodated by its new master.

Ritchey is the first to admit that because he was such a strong proponent of solid propellant rockets, there were some "who wondered if this attitude would be adverse to the success of RMD." He insists otherwise, that his actions were only taken out of concern for its health. He points out that as a consequence of his recommendations, Thiokol provided "a generous infusion of capital into the areas of business opportunities in which RMD might operate in the future."

---

\* Volume 11, AAS History Series.

As for Bernard Pearlman, formerly in charge of components and service, he agrees with Seymour that RMD was "a different animal." The Division was, in his words, "...essentially an R&D mechanical engineering and design company which was purchased by a chemical company..." He felt it a "fundamental mistake" to have placed RMD under the charge of individuals whose backgrounds were essentially chemical.

And so the arguments went.

## DIVISION MANAGEMENT

For the first couple of years after the merger, former RMI president Raymond W. Young served as General Manager. When, toward the end of 1960 he was given responsibility for the formulation and guidance of technical programs as Division Director of Advanced Planning, Seymour took over and ran Reaction Motors for most the the 1960 decade. He had joined RMI back in 1956 as Manager of Preliminary Design and later moved up to become Director of Research. Before RMI, he had worked with industry and the Office of Naval Research. When, in 1967, he was assigned to corporate staff, Lowell "Johnny" Matthies was named General Manager. He remained in charge of the Division until resigning a year or so later. Harry A. Koch followed in the spring of 1968.

Among the principal personnel at RMD during Seymour's regime were Richard Frazee, Director of Sales and Service; Harry A. Koch, Directory of Programs; Joseph Matolina, Director of Quality Control; Arthur Sherman, Director of Preliminary Design; Dr. Murray I. Cohen, Manager of Chemistry; Harold Davies, Manager of the Project Engineering Department; DeLacy F. Ferris, Director of Operations; Edward C. Govignon, Director of Programs; David J. Mann, Director of Research; Bernard Pearlman, Manager of the Components and Service Department; Albert G. Thatcher, Director of Product Planning; Charles Wilde, Manager of the Test Department; Hans G. Wolfhard, Manager of the Physics Department; William R. Wright, Manager of the Patents Department; and Theodore Zytkowicz, Controller.

At the time of the transition, Reaction Motors was in the process of changing from functional management to program management. Seymour recalls that "We made the move part way to staff program management and that didn't work very well. The trouble was that we were just not big enough to give our programs all the people and tools that were needed, so we remained partly function and partly program-oriented. What finally evolved was that I had two people reporting to me, Lace Ferris as Director of Operations and Ed Govignon as Director of Programs. Shortly afterwards, Lace left and Bill [William E.] Fogarty became Director of Operations."

In actual fact, it appears that policy differences led to Ferris' departure from RMD. He moved to North American Aviation and was assigned to work on an alternate propulsion system for the Condor program (see below). Later, the Navy canceled RMD's propulsion system in favor of NAA's. "Then," recalls Pearlman, ". . .Lace was sent back to RMD to take over whatever Condor tooling and test equipment he felt would be useful for his program. Can you savor the sweet feeling of revenge Lace must have felt. . . ?"

**Figure 1** General managers of the Reaction Motors Division, Raymond W. Young (top left), Edward H Seymour (top right), Lowell F. ("Johnny") Matties (bottom left), and Harry A. Kock (bottom right).

During the first few years that he headed up the Reaction Motors Division, Seymour reported directly to Ritchey whose Rocket Operations Center was headquartered in Ogden, Utah. When Ritchey became President in 1964 and moved to corporate headquarters in Bristol, Pennsylvania, Seymour reported to Joseph W. Wiggins, Vice President of Thiokol's Aerospace Group, also located at corporate headquarters. Wiggins had earlier managed the company's Redstone Division in Huntsville, Alabama. The four general managers of RMD, Raymond W. Young, Edward H. Seymour, Lowell F. Matthies, and Harry A. Koch are pictured in Figure 1 and Thiokol vice president Harold W. Ritchey is seen in Figure 2.

**Figure 2** Harold W. Ritchey photographed in November 1959 when he was Vice President and Technical Director of the Thiokol Chemical Corporation and President of the American Rocket Society.

## FACILITIES

When the old RMI left its original facilities at Pompton Lakes in New Jersey, the Navy had supplied space at their Naval Air Rocket Test Station in Lake Denmark, adjoining Picatinny Arsenal. The company also rented buildings in Rockaway; and, a few years later with Navy support, built a plant in Denville (next to Rockaway). The company owned the land and office building, the Navy the adjoining land, laboratory and manufacturing buildings. Later, when RMI became RMD, Thiokol purchased these facilities. Figure 3 is an aerial view of RMD facilities at Denville taken in September 1969, and Figure 4 is the Division's sign at the main entrance. Ground-breaking ceremony, 14 February 1964, for new Plant Engineering Building in Denville, is shown in Figure 5. Supervisor Plant Engineering Operations.

**Figure 3** Aerial view of the facilities of Reaction Motors Division, Denville, New Jersey, September 1969. The outline of the plant is clearly indicated.

**Figure 4** Sign announcing RMD at the Division's main entrance, Denville, New Jersey.

**Figure 5** Ground-breaking ceremony, 14 February 1964, for new Plant Engineering Building in Denville. Left to right: Bernard R. Faber, RMD Engineer and member of Denville Township Industrial Committee. Albert List, RMD Manager of Plant Engineering, Stanley P. Schmidt, RMD Director of Adminstration, Dr. Seymour, Mayor Foster, Theodore J. Zytkowicz, Howard Milligan, Chairman of Township Industrial Committee, and Herbert Clothier, RMD Section Supervisor, Plant Engineering Operations.

At first, most RMD production was assigned to the Reaction Motors Production Plant in Bristol, Pennsylvania. This facility was managed by E. Dana Gibson and later by Gerald M. Pacella. According to Harold Davies, Ralph Hoetger did a masterly job of industrial engineering. Other key manufacturing personnel were Fred Pfotenhaur, Edward Parke, and Laurance M. Levy, who for a period was manager of the division's purchasing department. The Bullpup A rocket engine (see below) production line was set up at Bristol; the Rockaway facility--though extensive and well-equipped--was a typically experimental or model shop and not easily converted to series production. When it subsequently came time to produce the larger Bullpup B engines, production did take place in Rockaway--but in a building near the ones previously occupied. All other experimental, component and miscellaneous work was assigned to Denville. Each of these three manufacturing facilities was approximately equal in size and manpower.

## DIVISION CAPABILITIES

Thiokol inherited from Reaction Motors a number of on-going projects and imaginative concepts in rocket engine systems and components plus a strong research capability. It is this latter in particular that Seymour felt was not appreciated by his new corporate superiors. He described for the author the research strength of the Division in the following terms:

> "There was a lot of very solid research done in chemistry under the leadership of Dave [David J.] Mann, [Dr.] Murray [I.] Cohen, and [Dr.] Stan[ly] Tannenbaum. RMD was one of the leaders in the country--perhaps world--in boron hydride chemistry, such as diborane and pentaborane. The discovery of vinyl decarborane opened up another whole field, offering promise of a highly energetic solid propellant binder. The discovery of a new polymer, nitrosofluoroamine provided an elastomer highly resistant to exposure to nitrogen tetroxide, the only one of its kind, thus invaluable for seals, O-rings, etc. And there were many others. A lot of this work was done under research contracts with the services, and some on company funds. The Physics Department under Hans [G.] Wolfhard also did interesting work, including some of the early work on electrically charged particle propulsion, a relative of the ion rocket concept."

As part of the pre- and post-merger process, Thiokol management made a thorough study of these and other Reaction Motors' capabilities. As a result, Ritchey elected to focus corporate support of division-developed liquid propellant rocket systems where: (1) the low-gravity environment and high specific impulse yielded an advantage over other approaches, and (2) multiple on-off and/or thrust-level variations were required. As pointed out by Seymour, "The ability to modulate thrust had always been a strong point in liquid propellant systems, as opposed to solid, where you can pre-program thrust variations by grain design, but on-command thrust control systems for solids were never too successful."

Some of Ritchey's colleagues suggested that a separate new entity be established within Thiokol to produce such engines, but he was not convinced by arguments to the effect that overhead would be reduced and profits increased on fixed-price contracts. Rather, he felt that the large sales base offered by Thiokol would itself help reduce RMD overhead and make the division more competitive.

At the time Thiokol acquired Reaction Motors, the development of the XLR-99 (later, LR-99) powerplant for the X-15 high-altitude, high-speed research aircraft was nearing completion. At the same time, it was becoming clear that testing this long-duration, comparatively high-thrust (50,000 pounds or 22,700 kilograms or 222 kN)) engine could not continue indefinitely at the Lake Denmark test area in densely populated northern New Jersey. "The noise, with 3-minute runs, must have been annoying to nearby residents," recalls Seymour. "We tried to cooperate, but with development behind schedule, we had to run whenever the engine was ready--which sometimes meant at night. To my knowledge, no physical (structural) damage to dwellings, etc. was ever proven. But it became obvious with residential build-up, that our test location days for large engines were limited." Figure 6 is a composite view of RMD's R-2 test area. The principal stand was designed to support the development of rocket engines producing up to 1 million pounds of thrust. Preparations are being made in Figure 7 to static-test an LR-99 rocket engine.

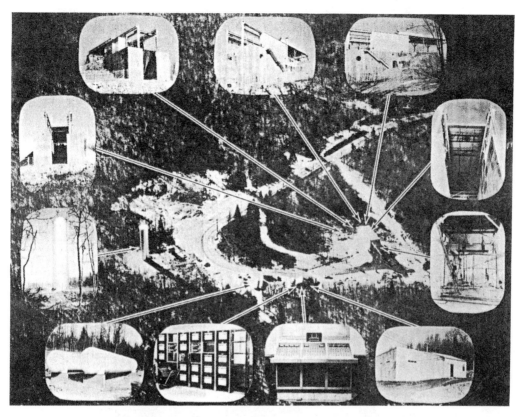

**Figure 6** Composite view of RMD's test area.

Already, legal proceedings had unfolded and although no punitive damages had been awarded by the courts, Thiokol realized that their new division could not go on developing and testing large rocket engines unless a move were made to a remote locality. In spite of this reality, management refused to authorize funds for such a move (which would have been very costly), feeling that the man-rated rocket field was already adequately served by larger and better-financed contractors. Even before the X-15 made its first flight, the future of RMD was in doubt.

Taking into consideration the realities of geography, Thiokol management assessed RMD's potential in terms of the following four principal pursuits:

1. *Packaged liquid-propellant rocket engines.* Used in the Bullpup A and B and other missiles, these engines would prove to be a productive venture, generating some $100 million in sales over RMD's lifetime.

2. *Small attitude control and maneuvering rocket engines.* RMD would enjoy limited success in developing and manufacturing vernier rockets for spacecraft, most notably for the Surveyor lunar soft-lander.

3. *Technologies and facilities applied to non-rocket propulsion endeavors.* Here, too, some success would result in such areas as steam generation and component development.

4. *Overall research and development.* It was hoped that company-sponsored R&D would open up new product opportunities for RMD as well as for the Thiokol enterprise as a whole. In Seymour's view, there was a lot of potential in this area that was never fully realized. As management attention focused on meeting aerospace obligations, product diversification and technology transfer languished.

**Figure 7** Test Stand R-3 used to test the LR-99 engine. The sloping tank at left is one of two propellant tanks that are integral to the stand. The large tank in the background across the road contains 150,000 gallons of water, enough to supply the entire test area.

## PROJECT DEVELOPMENT

In the paragraphs that follow, short summaries are provided of the principal projects and activities undertaken by Thiokol's Reaction Motors Division. These descriptions are in Frank H. Winter's article that follows.

### Propulsion for X-15 Research Airplane

Development of the LR-99 engine was about complete when Thiokol acquired Reaction Motors, though some problems persisted in the throttling system. The heat transfer out of the combustion chamber was nearly constant at all thrust levels, but at 30 percent of maximum thrust the amount of propellant to carry away the heat became marginal. The job of developing a fully throttlable, man-rated, fail-safe engine was greatly underestimated by all concerned, and work proceeded slowly until the orbiting of the first Soviet Sputnik in October 1957. With it providing the impetus, work accelerated and despite development difficulties, by 1959, Program Manager Harry A. Koch and Project Engineer Robert W. Seaman considered the engine to be developed and ready to turn over to NASA and the Air Force for flight testing.

But North American, the X-15's prime contractor, had completed the first airplane in mid-October of the preceding year. Since the LR-99 was not then ready, initial flight testing was accomplished with a pair of four-chamber LR-11 engines of the type earlier used in the X-1 program. Testing with these engines continued to X-15 flight No. 26 when, on 15 November 1960, Scott Crossfield flew plane No. 2 with the LR-99 installed. In all, three X-15s were built and flown for 199 missions. Twelve pilots flew nearly 46 hours at supersonic speed, including an hour above Mach 5. On 22 August 1963, Joseph Walker set an altitude record of slightly more than 67 miles (about 108 km). On another flight, a speed record of Mach 6.7 was reached. RMD received the NASA-Air Force Trophy for its contribution to the X-15 program.

The shipment from RMD of the first XLR-99 rocket engine is shown in Figure 8 and assembly operations are viewed in Figure 9. Figure 10 shows X-15/LR-99 participants celebrating the airplane's 100th flight at Edwards Air Force Base in California on 21 January 1964.

**Figure 8**  Shipment from RMD of the first XLR-99 rocket engine.

**Figure 9** Assembly of an LR-99 rocket engine by assemblymen and inspectors during a second shift at RMD.

**Figure 10** Celebration of the 100th flight of the X-15 at Edwards Air Force Base, California, 21 January 1964. Left to right: Harry A. Koch, RMD LR-99 project manager; Harold Buck, field service at EAFB; Bud Parker, Thiokol Aerospace Center (formerly RMD); Bill Arnold, field service at EAFB; Bob Seaman, project engineer on LR-99 and later RMD chief engineer, Jim Walker, RMD field service manager; Bob Bradford, Thiokol technical representative at EAFB, and Ed Seymour, RMD general manager.

There is an interesting footnote to the X-15 program. Back in 1960, G. R. Cramer and H. A. Barton proposed--at the National Aeronautics Meetings of the Society of Automotive Engineers in New York City--a means of sending the X-15 into low Earth orbit. They suggested that the plane be ground-launched by a powerful booster rocket. To build up the required orbital velocity, auxiliary droppable fuel tanks would be fitted to the X-15 and the engine would be driven by more energetic propellants than currently used.

In the same year, relates M. E. Parker, "the propulsion specialists at the Marshall Space Flight Center studied the LR-99 design features and control system which made the system fail-safe and man-rated, and later applied this knowledge to the design of the Saturn engine where man-rating was to be all important for the upcoming Apollo program."

## **Packaged Liquid Rocket Engines for Small Missiles**

Packaged liquid rocket engines consist of liquid propellants and a propellant pressurizing medium permanently sealed into a tankage shell that is integral with the rocket thrust chamber. In the early 1950s, RMI had recognized that such engines would incorporate many of the advantages of liquid propellants but could be handled in the field in much the same way as solid propellant rockets. Design studies by Arthur Seaman, Harold Davies and others and in later years supervised by Alan Maier, program manager for packaged liquids, indicated that a bi-propellant combination using a liquid oxidizer and a liquid fuel would be feasible depending on the solution of three problems:

1. The selection of a propellant combination and tankage construction material suitable for long-term storage, hermetically sealed.
2. The choice of a compact energy source to pressurize and expel the liquid propellants.
3. The choice of an injector and igniter device that, while being robust and readily producible, would enable a high performance to be achieved.

A basic engine concept eventually evolved that featured inhibited red fuming nitric acid as the oxidizer and an amine base fuel stored in an aluminum tank shell. For pressurization, a solid propellant charge was chosen; and, for ignition, a simple injector with igniting means was devised. To make the unit operational, all that was required was the activation of the igniter cartridge by the provision of electrical energy.

Reaction Motors Division executives, Figure 11, discuss storable propellant problems with RMI founder Lovell Lawrence.

Figure 11  RMD executives discuss storable propellants with Reaction Motors, Inc. founder Lovell Lawrence. Left to right: Albert G. Thatcher, Lawrence, E.B. Parke, Maurice E. Parker, and Richard Frazee.

## **Sparrow III**

An RMI 5,000-pound thrust (2.22 kN), 2.5-second duration unit was successfully test-fired in the mid-1950s that led to the company's receiving a request for proposal from the Navy's Bureau of Aeronautics for a packaged unit for the Sparrow III. Development of what became known as the LR 44-RM-2 or Guardian I began in December 1956 and in August 1958--four months after RMI's merger with Thiokol--initial production got under way. A problem arose in that the original Navy specification for the engine was 9,000 pounds thrust (40 kN) about twice as much as the missile could manage without guidance problems. Subsequently, the thrust was reduced. As it turned out, the production run did not meet division expectations and the Navy eventually opted for a solid unit for the Raytheon-built missile. According to Charles [Chuck] Dimmick "...the LR-44 performed very well. However, Raytheon was having extreme guidance problems," which led to difficulties in relations between the airframe manufacturer and RMD. This is covered in more detail in the next article by Winter.

During the period that small liquid systems were being considered for such short-duration tactical air-to-air missiles as Sparrow III, solid motors were encountering severe problems in achieving operational limits below -40 degrees F (-40°C). Packaged liquids, however, were able to meet -65 degrees (-54°C) requirements; and, in fact, had been tested down to -75 degrees F (-59°C). But, by the time that packaged liquids had progressed to the production stage in the late 1950s and early 1960s, solid systems had evolved that met temperature specifications and, moreover, were enjoying somewhat superior performance. This development had an immediate effect on Guardian I procurement and portended down-the-road problems for follow-on projects. In turn, this, contributed to the ultimate demise of the division.

## The Bullpup

Meanwhile, in September 1958--a bare month after the Sparrow III engine production startup--The Bureau of Aeronautics contracted with Reaction Motors to develop a packaged unit for the Bullpup A missile. Though larger and more powerful (12,000 pounds thrust - 153 kN) than Guardian I, the new LR-58 or Guardian II engine was based on the same design philosophy and powered by a hypergolic RFNA-mixed amine propellant combination. The propellants were permanently stored in hermetically sealed tanks of high-strength aluminum alloy in a pressure shell of welded construction. A solid propellant gas generator provided pressurization.

Testing got under way beginning in December 1958 at the Naval Missile Center, Point Mugu, California. Guardian II was also used in the GAM-83, the Air Force designation for Bullpup A. A boost-glide air-to-surface weapon, it was optically tracked and radio controlled. Purchases of Guardian IIs by the Navy and Air Force totaled over 33,000 units by 1964, when production was phased out.

Bullpup B was a larger version of the missile with a heavier warhead (1,000 pounds/454 kg versus only 250 pounds/113 kg for Bullpup A). The same guidance and control system served both missiles. In 1960, the Bureau of Naval Weapons awarded RMD a $2 million contract to design, test and manufacture a larger engine--the LR-62 or Guardian III--to power Bullpup B. By 1968, some 17,000 units had been built.

Bullpup A's engine was produced in an old World War II shell-loading plant at Bristol, Pennsylvania, originally owned by Charles Hunter and purchased by Thiokol for its new corporate headquarters around 1960. The Bullpup B engine line was set up at two rented buildings in Rockaway, New Jersey. For safety reasons, "fill and load" operations were performed at RMD's test area in nearby Lake Denmark.

Such large production runs were quite naturally accompanied by some problems. According to Seymour, after about a year or so of shipments, reports started coming back from the field complaining of leakage of the fill plug for the fuel tank. The problem was first noticed by Navy personnel in handling missiles on the West Coast. "We checked around," according to Seymour, "and found that it had also been noticed in other locations." What had happened was that after propellant loading into the Bullpup, the plugs had been sealed with a polymer. As it turned out, the storable fuel was a plasticiser for that particular polymer and simply ate its way through, causing the leakage.

RMD came up with a fix and over the next year or so, field service crews serviced 7,000 missiles from Germany to Japan and straightened the problem out. Meanwhile, all new missiles were fitted with welded seals.

The Bullpup and its propulsion system are seen in various views in Figures 12, 13, 14, and 15.

**Figure 12** U.S. Navy aviation ordnance technicians loading a Bullpup air-to-surface missile onto the wing of an FJ (Fury" jet fighter at the Naval Air Missile Test Center, Point Mugu, California.

**Figure 13** Bullpup engines LR-58, left, and LR-62, right.

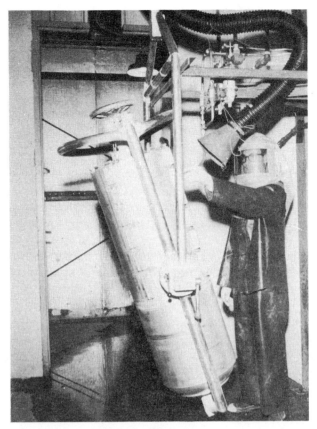

**Figure 14**  Filling the oxidizer tank of the LR-62 engine.

**Figure 15**  Final assembly of the LR-62 Bullpup engine, Rockaway, New Jersey.

## Corvus Air-to-Surface Missile

In 1956, before the Thiokol--Reactions Motors merger in the Spring of 1958, the U.S. Navy had contracted with Temco Aircraft Corp. to develop the Corvus air-to-surface missile. Soon afterwards, Reaction Motors received a subcontract to develop what became known as the Patriot pump-fed storable liquid engine. Despite progress made on the program under Thomas C. Tarbox's direction, the Navy announced on 18 July 1960 that Corvus had been canceled after the expenditure of $80 million since 1957. It was reported at the time that up to $450 million had been budgeted for further development and production. All the Navy would say was the "Corvus was more limited in its application than other systems now under development."

## Missile A/Missile B/Lance Activities

To capitalize on its experience with packaged units and to expand their utilization, RMD worked hard to sell the throttling capability inherent in liquids. According to Parker, one interested party was found in 1958 at the Army Ballistic Missile Agency in Huntsville, Alabama where work was progressing on a missile guidance scheme that required fully modulated, on-demand thrust variation to precisely offset drag encountered along the ballistic trajectory. ABMA, supported heavily by RMD with hardware and liquid expertise, demonstrated the arrangement with the Missile A program. A direct result was the selection of this guidance scheme and liquid propulsion for the battlefield Missile B, which became known as Lance. Competition was fierce for the development contract, which finally went to the Ling-Temco-Vought Aerospace Corporation (LTV) in 1962.

"Unfortunately for RMD," relates Parker, "LTV's arrangement used tankage which, although it used aluminum forgings and welding for the hermetic sealing and propellant storage à la Bullpup, was separate from what LTV called the 'engine,' which eventually was produced for them by Rocketdyne. Failure to win this program was a jolt to RMD's packaged liquid future, but undaunted, reinforced their resolve to keep it going as they turned to pursuing Condor."

## Condor Air-to-Air Missile

Building on earlier successes in the packaged liquid rocket field, RMD energetically pushed forward on a multi-million dollar contract to develop an advanced engine to power the Navy's Condor AGM-53A air-to-surface missile. Condor was originally conceived with a high-altitude boost and high-altitude cruise, but later the program was changed to a low-altitude cruise. The contract was awarded in 1966 by North American Aviation, Inc.'s Columbus Division. In the minds of some, the engine involved the "ultimate" in exotic propellants, new packaging technology, and, in Ritchey's words, "an appeal to anyone who wished to see a new rocket flame color--green." Condor was, in his view, "the last gasp of the original innovators of rocket technology--the first in, and now to be the first out."

Seymour disagreed that supporting Condor was a technically ill-founded move on the part of RMD. "Corporate management," he stressed, "was impressed enough by our early studies to invest a couple of hundred thousand dollars in a new test stand that embodied scrubbing chambers to remove the hydrogen fluoride (HF) from the exhaust gases." (The dissolved HF went to an impervious hold lagoon, where it was neutralized with lime, precipitating calcium fluoride, which was hauled away).

The Condor engine used chlorine trifluoride and mixed amine fuel an exotic combination that gave a reasonably higher specific impulse than more conventional propellants. However, the oxidizer was highly corrosive and therefore difficult to handle. But this was not the principal problem, as pointed out by Harold Davies. "The real problem was the amazing ability of chlorine trifluoride at raised temperatures to combine vigorously with anything in sight; fuel, chamber walls, injector or what have you. In a packaged liquid, it presented the worst of both worlds, not only a very difficult injector and thrust chamber design, but the need to completely isolate the oxidizer from the hot pressurizing gas. Although the specific impulse was high, the real performance advantages came in density impulse due to the high specific gravity of the oxidizer and a favorable O/F [oxidizer to fuel] ratio." Largely because of propellant problems, the missile never got out of the development stage; the Navy finally gave up on the liquid approach, and in the autumn of 1967 transferred development of the propulsion system to North American Aviation's Solid Rocket Division.

Davies recalls a vignette related to Condor. "When the program was changed to low-altitude cruise, my thoughts went to all of that good oxidizer going to waste and flying by the missile. I wondered if a much more benign system could be designed using a combined rocket-air breather. I chose a packaged liquid booster arrangement with a simple air-breathing sustaining chamber. The propellants were IRFNA [inhibited red fuming nitric acid] turpentine and 15 percent UDMH [unsymmetrical dimethylhydrazine] for the rocket boost with the turpentine mixture for the air-breathing chamber. Solid gas pressurizing and explosive valves were visualized. The air breather used a single stage centrifugal blower of 3:1 compression ratio driven by a blast turbine (à la early RMI experience) overlapping both the boost chamber and the sustainer. I sized the components from my gas turbine experience and drew a full scale sketch of the system. Even with the very modest performance chosen for the elements, it was evident that such a scheme would double the total impulse available within the Condor weight and volume limitations. I took my idea to Art Sherman, then in preliminary design, and wondered if we should be making an end run. He recognized it for the dynamite it was, and told me to do nothing and say nothing. He may have discussed it with others, but I am not sure."

### Supporting Work

Closely related to the Condor effort was the APPLE (Advanced Propulsion Packaged Liquid Engine) research program. Established to extend the technology of packaged liquid engines for a new generation of missile, it focused on the CTF (chlorine trifluoride) oxidizer and various hydrazine-based fuels. The ultimate aim of APPLE was the successful operation of an integrated single-chamber packaged liquid engine at two levels of thrust, with command restart of the sustained flight phase. APPLE was followed by a contract from the Bureau of Naval Weapons for a one-year effort to develop a high-performance Applied Research Engine (ARE) that was not to be designed to meet specific flight weight specifications. Still another mid-1960s activity was a contract with the Air Force Rocket Propulsion Laboratory to evaluate major parameters affecting the performance and durability of a flight-weight Radially Distributed Annular Rocket Combuster or RADARC. This design was to incorporate a relatively short combustion chamber that would deliver the same specific impulse as conventional cylindrical types, while permitting more propellant to be loaded into a given powerplant envelope.

In summing up RMD's experience in storable engines used in Bullpup and other applications, Harold Davies concluded that, "The real genius of the RMD system was the extreme simplicity and the direct coupling of the design concepts with advanced manufacturing techniques in welding of high-strength aluminum alloys. The production lines at both Bristol [Pennsylvania] and Rockaway [New Jersey] earned generous praise from such a cost-conscious diehard as Tom Willey, the General Manager of Martin Orlando [in Florida], then making the Bullpup. To produce loaded rocket engines for seven dollars a pound [$15/kg] was no small achievement. The demonstrated reliability of 0.997 is probably the highest number ever achieved by a liquid rocket system and equal to the best that a solid could offer. That these units [in the Bullpup application] stayed in service for 20 years should be noted."

Despite success with Bullpup, the failure of research projects to lead into funded programs and the cancellation of Condor were demoralizing to the division. "In my opinion," Seymour reminisced, "we let ourselves get too narrowly market-based. We never got into big engines. We were in small engines for tactical missiles. . . and small verniers for space applications. . . work on tactical missile development was grinding down to almost zero and Condor was the last such missile that fell within our market capabilities." Without Lance, without Condor, with Bullpup production coming to an end, and with such activities as APPLE, ARE and RADARC failing to generate hardware orders, the situation facing RMD was becoming worrisome.

Harold Davies elaborates: "My understanding was that in packaged liquids, at least, the Corporation [Thiokol] excluded us from pursuing that field [e.g. large engine development]. Some six months before Thiokol came aboard, I proposed to Lace Ferris that RMI fund a packaged liquid work-horse of around fifty thousand pounds thrust [222 kN] with a running time of eight to ten seconds. IRFNA-mixed amine propellants. I suggested that it could be done for $25,000.

"At first he was skeptical," Davies continued, "but when I assured him that we intended to have a true packaged liquid with cross slide initiation, he accepted the idea with enthusiasm. We constructed a steel workhorse using one of the old Bomarc ceramic lined thrust chamber (450 psi) [3,100 kN/m$^2$]. Hercules at Kenvil agreed to make six pressurizing grains in the reliable OGK double-base propellant for around $4,000. These were about 8 inches [20 cm] in diameter by 4 feet [1.22 m] long, and were to be delivered without testing by them or indeed by us!

The hardware was completed shortly after the Thiokol merger. We were told that we could make one firing only and that there were to be no more under any circumstances. Lee Bachman made the shot and it was quite successful. . . Later, it did assist in convincing the Navy that the Bullpup B scale-up would be reasonable. In these decisions, however, there was the strong assumption that RMD would not be permitted to do anything that might impinge, even remotely, on the large solid booster. No doubt this restriction would not have applied to large pump-fed systems. I do not recall that the restriction was ever in print but I believe it was well understood."

## Small Vernier and Attitude Control Rockets

Despite this, one area did appear to hold promise. According to Ritchey, a respectable market had appeared for small control motors for both spacecraft and military missiles. However, he felt that "RMD seemed to be out of the main stream, for reasons that are hard for me to understand." One reason for the division's lack of staying power in the vernier field may have been competition from the Marquardt Corporation, which had developed successful motors that utilized superior high-temperature materials.

Despite ultimate disappointments in this technical area, RMD did successfully develop liquid propellant vernier rockets for the seven Surveyor lunar soft-landing craft flown by the National Aeronautics and Space Administration (NASA) between May 1966 and January 1968. A small, throttlable liquid-propellant engine was developed by RMD for mounting in sets of three on Surveyor to make midcourse corrections between the Earth and Moon; and, during the lunar landing maneuver, to trim descent velocity after the spacecraft's big retrorocket (made by Thiokol's Elkton Division) had fired and also to assure a level landing.

The engine, designated TD-339, was designed with the special requirements that it be throttled independently for pitch and yaw control and be gimbaled for roll control. Thrust output, which varied from 30 to 104 pounds (13-463 N) and could respond to multiple in-flight restarts on command, had to be regulated to control the landing velocity, which was to be sensed by an altitude radar. The TD-339 consisted principally of a thrust chamber and injector assembly, dual-propellant throttle valve and shutoff valve. The engine weighed about six pounds and measured 11 inches in length.

Total thrust level was controlled by an accelerometer at a constant acceleration equal to 0.1 Earth gravity. Pointing errors were sensed by gyros, which could cause the individual engines to change thrust level to correct pitch and yaw errors and swivel one engine to correct roll errors. Gold plating on the external surface of the thrust chamber maintained the proper passive thermal balance, preventing propellant freezing or boiling during the Earth-Moon trajectory.

At approximately 100 kilometers slant range from the lunar surface, the marking radar started the flight control programmer clock, which then counted down a previously stored delay time and subsequently commanded ignition of the solid propellant main retro-rocket along with the three TD-339 verniers. The latter maintained a constant spacecraft altitude during main retro-firing in the same manner as during mid-course thrusting. After main engine burnout, the flight control programmer controlled the three verniers' thrust level until the radar altimeter and Doppler velocity sensor (RADVS) locked up on its return signals from the surface below. Descent was then controlled by the RADVS, whose signals were processed by flight control electronics to throttle the verniers. At four meters altitude, when Surveyor would be traveling at less than two kilometers per hour, all three verniers were cut off and the spacecraft fell freely to the ground.

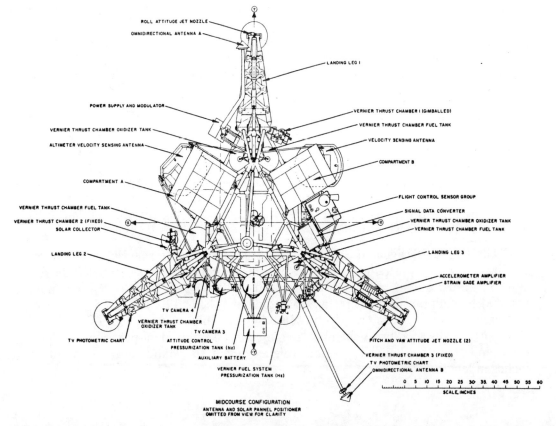

**Figure 16** A drawing of the Surveyor spacecraft showing the location of the RMD vernier units.

While the Surveyor vernier program was finally successful, there was a period when it seemed that RMD might not be involved to its conclusion. Seymour remembers receiving a call one night from Robert Roney, chief scientist of Hughes Aircraft, Surveyor's prime contractor to the Jet Propulsion Laboratory. "'Ed, you're going to get this telegram tomorrow terminating the contract!" He went on to add that "I just want to let you know ahead of time." I said to myself, there really is something to this Friday the 13th [of March 1964]. And on Saturday, the 14th, we officially got the word of the cancellation."

Seymour immediately hit the road and visited JPL and Hughes trying to determine what had gone wrong. Back on the East Coast, RMD was busily changing over to more effective program management and Seymour so informed Gene Giberson, JPL's program manager. From then on, the situation improved and RMD was reinstated in the Surveyor program under direct contract with JPL.

It seems that Hughes was having troubles with most of their subsystems contractors, and was looking to North American or Space Technology Laboratories as possible replacements for RMD. But neither could meet the schedules JPL had established for the program; so, five weeks after the cancellation, Seymour got another call from JPL: "Hey, Edward, would you consider taking the job back?' And I said, "Just give me time to think--I had taken a gamble and had kept the Surveyor vernier crew intact--30 seconds will be enough!"

Figures 16, 17 and 18 show the Surveyor spacecraft in drawing and actuality. Its terminal sequence is depicted in Figure 19, beginning at about 1,000 miles above the lunar surface. The main retro burns out in 40 seconds at about 25,000 feet altitude, reducing the velocity to approximately 250 mile per hour. After burnout, the flight control programmer adjusts the thrust levels of the RMD vernier units until the Radar Altimeter and Doppler Velocity Sensor, or RADVS, locks up on its return signals from the Moon's surface. Descent is then controlled by RADVS and the verniers. At about 13 feet, velocity is slowed to some 3 miles per hour and the verniers shut off. The spacecraft then free-falls to the ground.

## C-1 Common Engine

Designated TD-345 C-1 and referred to both as the common and the Radiamic engine, it was a small, 100-pound (45-kilogram or 445 N) thrust unit designed for extremely precise applications. Ray Novotny was principally responsible for its design. C-1 was called *common* because it was considered for a number of different NASA space vehicles, and *Radiamic* because, as seen in Figure 20, it employed regenerative-radiative cooling that exhibited heat transfer properties substantially lower than any purely radiative, regenerative, or ablative system in the same thrust class. Provision was made in the C-1 for a detachable nozzle extension to allow substitution of different exit cones, so that the same basic chamber could be adapted to various attitude control, maneuvering and ullage system installations.

**Figure 17** Surveyor spacecraft being checked in the contractor's facility prior to being shipped to the Kennedy Space Center in Florida. Note the vernier thrust chamber at the right.

**Figure 18** Underside view of the Surveyor providing a clear view of the verniers.

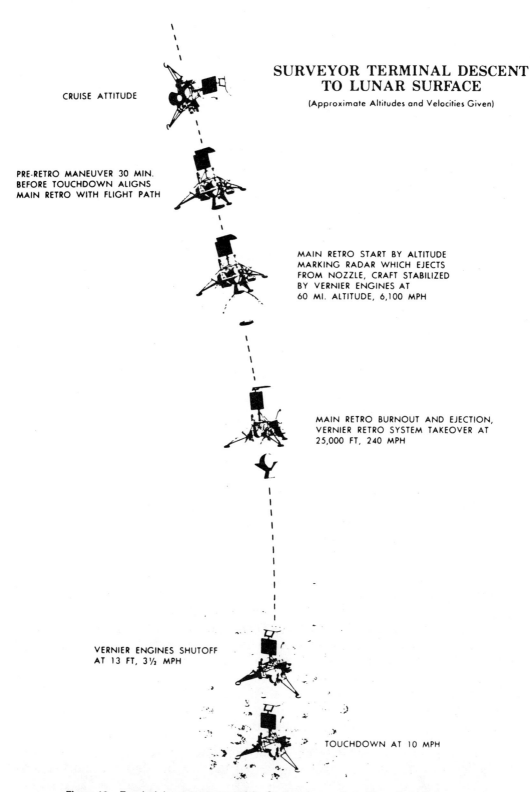

**Figure 19** Terminal descent sequence of the Surveyor from ending of its cruise attitude to touchdown on the lunar surface.

**Figure 20**  TD-345 C-1 Radiamic unit, the so-called "common" engine.

C-1 was a significant advance over the Surveyor vernier with the difference that its ability to be throttled was to be achieved by pulsing rather than by modulating propellant flow. C-1 applications for which the engine was initially designed to fit existing envelopes, propellants, environments, and other requirements were: Apollo command module, service module, and lunar module; Saturn S-4B ullage; and Gemini. All of these applications used nitrogen tetroxide as the oxidizer and one of the following hydrazine fuels: monomethylhydrazine, unsymmetrical dimethylhydrazine, neat hydrazine, or some blend of these. RMD received a $1.5 million contract in 1965 from NASA's Marshall Space Flight Center to develop the engine.

To Seymour, the C-1 development was fascinating. "The engine design and construction were not too tough," he told the author; "the big problem was throttling and simulating space flight profiles." Specifications called for reproducible pulses from a few milliseconds to several minutes under space conditions. The C-1 program resulted in heavy company investment in an altitude test stand, completely computer-controlled. Simulated flight profiles could last for many hours, Seymour reported. The computer would process the test data, permitting the test conductor to receive reduced data on a real-time basis. Seymour felt the test operation was probably one of the most advanced in the country at that time.

Although the C-1 ended up fully flight-qualified, it never became operational. Parker felt that "the development was started too late by Marshall and failed to overtake the primary motors that Johnson had selected and developed earlier at Marquardt and Rocketdyne." Nevertheless, RMD engineers felt that the chamber design they had conceived and developed at the 100-pound thrust (445 N) level could be adapted to families of large space and tactical propulsion systems delivering up to 10,000 pounds (44.5 kN) of thrust.

## The Nerva Nuclear Rocket Project

According to Ritchey, Edward Teller--a member of Thiokol's Technical Advisory Board--had recommended that RMD not compete in NASA's Nerva nuclear rocket effort "since it was doomed to failure and might damage our reputation." Nevertheless, Ritchey added, "We made an all-out effort to get the Nerva development, even though more of us thought that it would never fly. In spite of what I thought was an excellent proposal, Aerojet won the program. RMD desperately needed this addition to a rapidly dwindling backlog. We proposed doing the testing in Nevada, so that the noise problem would not have been an inhibition."

Some work on nuclear rocket design had been going on at Reaction Motors under the leadership of John Newgard even before the merger with Thiokol. Thus, there was talent on board when Nerva got under way in 1961. Seymour remembers that RMD sales chief William Davidson had presented the Division's nuclear capabilities to Thiokol's Technical Advisory Board in Ogden, Utah, without receiving much encouragement. Ultimately, RMD did obtain some component work associated with Nerva that lasted until NASA phased out its nuclear rocket activities in the early 1970s.

## Steam Generators

RMD moved into the steam generator business as a result of wanting to perform altitude-testing of its smaller engines. The normal way of doing this was to use stream air ejectors to lower the pressure to the desired simulated altitude; but, for RMD, the cost to install boiler capacity at its test area would have been excessive. So, recalls Seymour, engineers "came up with the idea of decomposing peroxide, adding fuel gas to burn with the oxygen and using this as the steam generator." Although RMD realized that operational costs might be fairly high, it was felt that the steam-generating reaction could be accomplished in a unit about four feet long instead of a boiler the size of a small house.

"So we built a set-up like this and took it over to the test stand," Seymour related. "A minor business soon got under way and over the next year or so we built steam-generating and ejector pump setups. . ." that were used by RMD itself, the Thiokol Elkton Division, and such clients as Jet Propulsion Laboratory, Space Technology Laboratories, Bell Aerosystems, Rocketdyne, White Sands and NASA-Marshall. The system was marketing under the name Hyprox.

In operation, the RMD generators were able to provide a high-capacity pumping system to remove gases from the testing chamber where a rocket motor was undergoing test, thus reducing the pressure at the test nozzle exit and simulating altitude. The generators produced up to 750,000 pounds (340,000 kg) of steam per hour at 1,400 degrees F ($760^{\circ}$C) and at pressure up to 150 psi (1000 kN/m$^2$).

## **Components**

"Components," remembers Bernard Pearlman (Figure 21) "were my special pleasure." He took over responsibility for their development in 1958 and in four years increased sales four-fold. His group developed and manufactured, for example, valves for the Air Force Atlas and Titan intercontinental ballistic missiles. For the former, they built many fast-acting shutoff valves for engine staging; and for the latter, valves for the propellant tanks pressurization system--a hermetically sealed, flapper-type unit.

**Figure 21** Bernard Pearlman, manager of the Components Department, and Product Sales manager Stanley Nemick, left and second from right, man the RMD booth at a November 1959 meeting of the American Rocket Society, Sheraton Park Hotel, Washington, D.C.

Other component work included 24 sets (144 individual units) of hermetically-sealed pre-valves for Titan II and Titan III and 12 vibration monitors for Saturn 5. RMD also received a contract to produce and supply the vibration monitors for Apollo service modules. Their purpose, to monitor combustion stability and vibration levels of the principal on-board propulsion system.

Pearlman was, for a period, assisted by William Fogarty as Program Manager for valve work, and by an extremely gifted design engineer, Zola Fox. Pearlman still voices keen disappointment that Seymour and Ritchey terminated the components R&D budget in 1962, which, he observes, "presaged the end of the components operation." He admits that he could ". . .never understand why they wanted to 'kill the goose that laid the golden eggs.' Components at that time, accounted for half the Division's profits and a quarter of overall sales."

## Other Projects

Perhaps the most intriguing of miscellaneous projects undertaken during RMD's 14-year lifetime was the resurrection of the old four-chamber LR11.[2] Seymour relates the story: "Just before I left RMD [in 1967], a guy named John McTigue, head of NASA's lifting body program, got in touch with me and said, 'You know, we have the solid retrorockets for landing but would like to stick in some more power for the transonic speed range; we're going through it too fast. I've searched around and either the Agena engine or your old LR11s would fit into my craft. And I did a little more digging and found that there are six LR11s on the base at Edwards. Will you guys give us a bid on refurbishing them, replacing the seals and so forth, and give us a set of plans and specifications'?

"Our guys [at RMD] were not at first enthusiastic about it; it wasn't a big job. I don't think we had a set of plans. But my view of this. Look, Dyna-Soar had just been killed and if we did something like this [what McTigue had proposed], it would keep us in touch with developments for future spacecraft. Well, we got the job, pulled out the old engines, and sent a man out there, Billy Arnold, as service engineer. Our old engines were flown in something called the HL-10, I believe, and others. . . So that one engine flew for a good 30 years."

As it turned out, the LR-11 was installed not only in the HL-10 (beginning late 1966), but also in the M-2 (beginning early 1967). The lifting body concept was at the time under investigation by NASA's Office of Advanced Research and Technology to help establish the technological base for design of future manned re-entry vehicles. HL-10 and M-2 lifting bodies were carried aloft under the wings of B-52 bombers to about 45,000 feet (13.7 km). Unpowered glide flights lasted about four minutes, powered flights about eight; and, the respective velocities were 350 (560 km/hr) and 1,000 miles per hour (1600 km/hr). All flights took place at Edwards Air Force Base in California.

## Other Activities

Over the years, RMD became involved in a variety of other activities, of which research into the field of propellant chemistry was probably the most important. As an example, Division chemists investigated the combination of oxygen difluoride ($OF_2$) and diborane ($B_2H_4$) and achieved vacuum specific impulse values of about 390 seconds. Testing was conducted at the Air Force Arnold Engineering Development Center in Tullahoma, Tennessee, at simulated altitudes in excess of 125,000 feet (38 km) using a 2,000 pound thrust (8.9 kN) 40:1 expansion ratio engine. Coupled with its high bulk density, $OF_2/B_2H_6$'s long-term storage capability eliminated substantial weight penalties and system complexities associated with cryogenic propellant boil-off losses and insulation requirements.

In another area, RMD developed the completely fluorinated elastomer Carboxy Nitroso Rubber (CNR) for aerospace and military use. Sheet stock samples of CNR were found not to burn at any pressure level. Because of its properties, NASA and its contractors on the Apollo program utilized the RMD product to flame-proof critical elements of the manned spacecraft (space suits, couch foams, electrical wire coatings, potting and encapsulating compounds, etc).

A project started in the RMI days and completed in 1959 shortly after the merger was the Internal Combustion Catapult Powerplant (ICCP). According to Pearlman, the project got started back in 1948 in RMI as a feasibility study to determine if it was possible to develop very high pressures in rocket chambers. The project floundered for a time, only to be revived later. The initial approach used liquid oxygen and gasoline which were injected into the chamber by a stepped piston pressurized by chamber gases. This concept was devised by Pearlman. The propellants were later changed to compressed air/diesel fuel and water for improved safety and reliability. It was a fully servo-controlled system with 70 million foot-pounds (95 MJ) total energy.

As Seymour explains, "The Navy's first nuclear carrier, *Enterprise*, was under construction, and there was some concern as to whether her type of propulsion plant would provide the steam surge capability for the then-new steam-powered, slotted-tube catapults. There was a sophisticated control system--the aim in any catapult is to maintain what the old steam engineers would have called a 'square indicator card' where pressure is recorded against piston travel. The point is that through the entire piston stroke you want to be delivering the maximum energy to the aircraft, without overstressing the structure.

"A prototype system was built and installed by the Navy at the Lakehurst Naval Air Station in New Jersey. The qualification program turned into a never-settled debate; to us, the Navy kept inserting changes rather than following the stated plan. But, as the arguments proceeded, the *Enterprise* steam problems were settled." Meanwhile, RMD had already delivered and installed a complete system on the aircraft carrier.

The reason the Navy was interested in the system, recalls Pearlman--who worked on the project and received a patent for the final system-- ". . .was that they projected much heavier and faster carrier aircraft which the steam catapult powerplant could not accommodate because the system could not maintain pressure. It was when our ICCP was installed on the carrier that the Navy realized that their earlier projections of aircraft speeds and weights would not materialize and hence the ICCP would not be needed."

An extensive program was established at Lakehurst to train naval personnel in the operation of the catapult, according to Chuck Dimmick. "Our Service Department never did anything on catapults," he explains. "Engineering thought they should run the show." Elaborated Harold Davies: ". . .its purpose was to fully try out a shipboard system under realistic conditions. It was to be a qualification of the system as well as an opportunity to train Naval personnel. Firings ran into the thousands."

The final example of miscellaneous programs undertaken by the Reaction Motors Division was the rocket belt, to Seymour, a "most unusual" project. He explains: "In the 1960s, we met with Army Engineers at Fort Belvoir. They were interested in advanced ways to move troops across rivers, ravines, etc. We came home and brainstormed this, deciding that, for a first step, we would go to the Buck Rogers approach. So we strapped a rig on a man, with a rocket nozzle off each shoulder, powered just by high-pressure nitrogen from a tank on his back.

"Our 'test pilot' didn't have much luck in maintaining attitude control, landing on his head frequently. So the nozzles were moved to the hips, and that permitted the human's remarkable body control to work--like tumblers. At this point, we lost our test pilot. He was a test area technician who'd had problems in his marriage. But he and his wife made up, and no way would she let him continue this activity.

"While ruminating on what to do next, we had a Saturday visit from the promoter of the Aquacade show at the New York World's Fair--then about 6 months away. He wanted a man to fly across the Aquacade pool--from diving platform to diving platform. He said, "If you're in doubt, I'll write you a check for $300,000, right now!' We fell out of our chairs."

The name of the promoter was, according to Chuck Dimmick, something like Dohorick. "RMD turned him over to me for lunch," he recalls--"three martinis at the Three Sisters and then one for the road. He asked me [in accented English] what RMD would do. I said they'll turn you down. He said 'who, where I get job done?' I said in someone's garage. "Who garage--you the only one seem to know what--you garage'?"

Seymour picks up the story. "Heavy discussions ensued. The problem was that the maneuver called for a 'hot' system; there wasn't enough energy in the 'cold gas' nitrogen system. We had not tested such systems, and knew from our X-15 experience what's involved in 'man-rating' a system. We didn't think that it could be done in time, and were highly conscious of what would happen if Thiokol's name was linked with a failure in such a setting. So we declined. Bell Aircraft picked the project up, and performed it beautifully."

## THE EVOLUTION OF DIVISION FORTUNES

Shortly after the merger with Thiokol in the spring of 1958, Reaction Motors managers requested approximately $4 million to invest in test facilities that would permit them to enter the liquid hydrogen rocket propulsion field. The amount of the request, coupled with the fact that competitors Pratt & Whitney, Rocketdyne, and Aerojet had been firmly established in this field for a number of years, led Thiokol management to turn down RMD's request.[3]

Failing thus to interest corporate management, RMD turned to small attitude control and maneuvering rockets; however, it was several years before the Division met with success. In the meantime, Bell, Rocketdyne, Aerojet, Marquardt, and TRW were all attempting to gain a dominant position in this area. With the winning of contracts in the early and mid-1960s for the Surveyor vernier and the C-1 common engines, it appeared that RMD might do well in the small rocket market.

Over the years following its acquisition, RMD received a series of capital infusions for facility buildup and other purposes to the extent that the cumulative cash flow (interest on investment and taxes considered) became negative by about $4 million toward the end of 1967. In addition, from 1958 through 1969, funding for research and development totaled $5.2 million, with an annual rate in excess of $500,000 for each of the final five years.

Reaction Motors Division sales built gradually from 1958, reaching $35.7 million in 1963. The figure dropped slightly to $33 million the following year, to $30.4 in 1965, and then down to $27.7 million in 1967. This was followed by a dramatic decline to $7 million in 1968, an occurrence directly attributable to the phasing out of the production of Bullpup B engines and to the expenditure of considerable company funds on the Condor. To some, this was an unwise investment; to others, a reasonable gamble to protect the future of a much larger equity.

At the beginning of 1967, RMD had over 1,400 employees who had been required to service the relatively high annual sales of the previous years. By the end of the year, four major programs were nearing completion or being phased out: Bullpup production, Surveyor vernier development and production, C-1 development, and Condor. No new programs offering significant financial impact could be discerned, a fact that led to a reduction in division employment to approximately 530 persons by the end of the year. Manpower requirements continued on a downward trend during 1968, forcing further personnel reductions to be carried out. Termination pay for 1967 and 1968 reached, respectively, nearly $800,000 and somewhat over $335,000.

As RMD's fortunes waned, a concerned Ritchey reported to the Board of Directors on the company's "Dog House Half Dozen," of which RMD was characterized as the worst member. His financial analysts were projecting division losses of $2.746 million for 1968. As it turned out, when all the figures were in, the operating loss for that year was $1.9 million. Though not as severe as projected, something, Ritchey warned, would have to be done and done quickly if the division were to survive.

After detailed review, discussion, and direction, plans for increasing sales and reducing losses were established as goals. A number of economies were introduced, personnel was further reduced, and efforts were made to achieve more efficient operation. Work on the Army's short-range Lance ballistic missile produced some much needed income, and, during July, August, September and October 1968, helped the Division show a small operating profit. Unfortunately, the contract was terminated shortly after. This, in turn, led to further reductions of RMD personnel coupled with attendant expenses and overall operational losses for the months of November and December.

The situation at the beginning of 1969 is summarized by the following figures:

Number of employees - 298

Number of active contracts - 33

Number of contracts completed but not closed out - 93

Contract backlog value, $ millions - 2.3

Negotiated contracts not received, $ millions - 2.1

Projected 1969 net sales, $ millions - 6.9

Net plant, property and equipment, $ millions - 6.139

Even though management projected net sales of $6.9 million for the year, RMD was faced with an estimated net operating loss of $1.5 million for the same period. Completely reliable projections could not be made, however, since the results of further personnel layoffs and overhead cost reduction efforts were not immediately available. Toward the end of January 1969, management completed a detailed manpower requirements study based on the latest contract projections. As a result, by the end of February, RMD began to reduce its manpower to 270.

For 1970, RMD submitted data to Thiokol's Board of Directors indicating sales of $10.387 million and an operating profit of $560,000. The board felt this to be an overly optimistic projection largely because $3.3 million in sales was based on a new Bullpup B missile production program. A follow-on contract to execute such production did not, in the board's view, appear likely.

The obvious solution to RMD's problems was to generate increased sales that, hopefully, would lead to profitable operations. On the basis that breakeven was approximately $2.5 million above projected sales, and since this was a relatively small amount when compared to total Thiokol Aerospace Group sales of $180 million, it might have been concluded that breakeven operation could be achieved. This, however, was not to be the case. For one thing, the Department of Defense was then in the process of making an across-the-board reductions in its research and development budget. Perhaps even more significant was the fact that funding for liquid rocket technology was being cut to a greater extent than for solids.

There thus seemed to be no realistic missile or spacecraft engine development program on the horizon for which RMD could successfully compete. The division did continue to obtain contracts in the liquid rocket component business covering research, development and manufacture. Also, in late 1968, it received a subcontract from the Chamberlain Manufacturing Corp. for machining 8-inch (20 cm) artillery shells. Chamberlain was operating contractor of the Scranton Army Ammunition Plant (in Pennsylvania) and was a major supplier of ordnance devices for the military. Initial production began in July 1969, with gradual buildup to 5,000 shells per month. Work was conducted at a newly established ordnance plant in Rockaway, New Jersey, utilizing government-owned equipment residual to the Bullpup program. At about the same time, RMD added a metal parts product line as it continued diversification as its Denville operation.

One interesting effort, recalls Seymour, was the sale of testing services to chemical companies for safety and stability of their product and process intermediates. This was based, he said, "on our experience built up over the years in testing new propellants to see whether they could be handled safely. Under Chet Grelecki [Dr. Charles J, former Manager of Research] and Steve[n] Thunkel, this was carried on as an independent business after RMD was gone." The company, Hazards Research, "is still viable and doing well" according to Chuck Dimmick.

## THE END

Such relatively minor successes notwithstanding, Thiokol's management continued to feel that RMD's chances of generating major new contracts were rather remote. When, at the beginning of 1970, the 1969 Annual Report was published, the observation was made: "Early in the year, a decision was made to discontinue our participation in the liquid propellant field, since we could not identify sufficient profitable business in the future to justify continuation. As a result, the activities of the Reaction Motors Division are being phased out."[4]

The report went on to point out that RMD had essentially completed work on its contracts by the end of 1969, that some surplus equipment had already been sold, and that parts of its facilities had been leased.

Thiokol management, in the meantime, had been wrestling with four potential alternatives: (1) sell RMD as a continuing business operation; (2) lease its facilities; (3) discontinue all operations and sell off assets; or (4) continue operations at reduced staffing and facility use so as to match closely division business potential (excess facilities, according to this scenario, would be leased).

When the Condor was terminated, in the autumn of 1967, the first alternative had been pursued. Several companies had been contacted in an effort to sell RMD. Among them were Fairchild Hiller, Marquardt, Beech Aircraft, and Bell Aerospace. Bell had expressed some interest; but, it subsequently indicated that company workload did not justify operating at two locations. Approach No. 1 was soon abandoned.

The second and third approaches were pursued simultaneously, with some buildings being temporarily leased and then sold. Meanwhile, buyers could not be found for the land under them which is still owned by Thiokol. Reaction Motors' Denville and Rockaway facilities later became industrial parks (Figures 22 and 23), and those at the Lake Denmark test area served as the site of a company working on irradiated food technology. Finally, the machinery and equipment in RMD buildings were auctioned off by a Detroit firm. With this, the Division came to an end as a tangible entity.

**Figure 22**  Reaction Motors Division's former main facility, Denville, New Jersey. Front Entrance is to right, manufacturing wing to left.

**Figure 23**  The former administration and engineering building in Denville, looking north.

In June 1972, some three years after the decision was made to suspend operations, 14 years after it was established as an element within the Thiokol umbrella, and over 30 years since the founding of Reaction Motors, Inc., Thiokol's Reaction Motors Division ceased to exist.

Asked by the author to recall the feelings experienced during and after the acquisition of Reaction Motors, Ritchey said that initially he was strongly opposed, then became more or less neutral and withdrew his opposition, and "for a short period" after the acquisition, experienced a period of "positive enthusiasm." To Seymour, "The RMD years were, in many ways, the most exciting, hair-raising, and satisfying years in my work career." He later added that "It's easy to second-guess, in hindsight. Things could have been done differently. For instance, RMD's chemical research team and Thiokol's Chemical Division research team could have been combined into a terrific innovative force. Corporation divisions precluded this. . . Suffice it to say that RMI/RMD was a bubble on the cascade of rapidly developing technology in America. But a pretty important bubble, and one with which I will always be proud to have been associated."

## REFERENCES

1. Frederick I. Ordway and Frank H. Winter, "Reaction Motors, Inc.: Corporate History, 1941-1958.", *AAS History Series*, Vol. 11, ed. M. R. Sharpe, 1991.

2. Frank H. Winter and Frederick I. Ordway, "Reaction Motors Inc.: Project History, 1941-1958.", *AAS History Series*, Vol. 11, ed. by M. R. Sharpe, 1991.

3. Much of the information on the financial evolution of the Reaction Motors Division is based on research at Thiokol's former headquarters in Newtown, Pennsylvania by Frederick I. Ordway on 15 February 1982, and discussions with Messrs. T. G. Tumelty, former Supervisor of Records Management, and Albert J. Roeper, former Vice President - Finance and Treasurer; an interview with Roeper by Ordway at Newtown on 9 May 1983; and various letters from and telephone conversations with Tumelty and Roeper in 1985 and 1986.

4. Use was made throughout the article of Thiokol *Annual Reports* from the late 1950s to the early 1970s; of Aerospace Facts published monthly by the Aerospace Center, Thiokol Chemical Corp., Ogden, Utah; of the monthly *RMD Rocket* newsletter, the weekly and special *Rocket Bulletin* and the *Rocket Weekly News Bulletin* published at RMD. Appreciation is expressed to Carson B. Trenor, Morton Thiokol, Inc., Chicago, Illinois and George Uibel, Morton Thiokol, Inc. Wasatch Division, Brigham City, Utah.

### Other Sources

Personal and telephone interviews were held with former members of the Reaction Motors Division supplemented by letters. Instead of providing reference numbers to each quoted source, listed below are the persons who helped make this article possible and supporting details on how they did it.

*Seymour*: Interview with Dr. Edward H. Seymour by Frederick I. Ordway on 25 January 1986, Buckingham, Pennsylvania, prior and subsequent telephone conversations, and letters and appended material from Seymour to Ordway dated 17 December 1985, 8 and 10 February 1986, and 22 April 1986; final telephone interview on 14 June 1986; and letter dated 27 June 1986.

*Ritchey*: Letters from Dr. Harold W. Ritchey to Frederick I. Ordway dated 20 December 1985 and 31 January 1986, and telephone interview on 14 June 1986.

*Bell*: Letter from Harold S. Bell, Jr, to Frederick I. Ordway dated 7 and 18 February and 3 June 1986; interview by Ordway in Madison, New Jersey, on 22 June 1986 and visit to former Reaction Motors' plant sites at Denville and Rockaway, New Jersey; and letter 6 September 1986.

*Dimmick*: Letter from Charles Dimmick to Frederick I. Ordway dated 3 June 1986 and earlier meetings with him at Lake Mohawk, New Jersey, in connection with Refs. 1-2 and applicable to the RMD story.

*Parker*: Interview with M. E. (Bud) Parker by Frederick I. Ordway at Thiokol's facilities at Redstone Arsenal, Huntsville, Alabama, on 27 April 1983, subsequent telephone interviews, and letter with attachments dated 22 June 1986.

*Davies*: Harold Davies, letters to Frederick I. Ordway dated 9 July and 21 August 1986.

*Pearlman*: Letters from Bernard Pearlman to Frederick I. Ordway dated 17 July and 18 August 1986.

AAS 91-292

Chapter 12

# REACTION MOTORS DIVISION OF THIOKOL CHEMICAL CORPORATION: A PROJECT HISTORY, 1958-1972 (PART III)[*]

### Frank H. Winter[†]

**SPARROW III**

The original Sparrow air-to-air rocket was one of the U.S.' oldest missile weapons, starting during World War II. It went through several versions, all solid-fueled. Solid-propellant powerplants were then the most practical means of propulsion for all air-to-air missiles. These weapons were invariably small and did not require complex liquid-fuel systems. Their ranges were short and propulsion requirements limited to high thrusts for short durations. Moreover, the missiles had to be prepared for firing in a moment's notice even after long-term storage, yet able to withstand wide temperature ranges during storage. Solid-fuel motors perfectly fit the bill for these criteria, though specific impulses were lower vis-à-vis liquid fuels.

However, during World War II the Germans did build the X-4 air-to-air weapon powered by a BMW (Bayerische Motoren Werke) bi-propellant liquid rocket motor. Simplicity in operation was achieved with hypergolic fuels of nitric acid and a hydrocarbon (code names, "Salbei" and Tonka"). That is, they ignited on contact, dispensing with the need for an ignition system. The propellants were forced into the combustion chamber by the sudden release of compressed air. This was initiated by the pilot pushing a single switch. The Germans thus realized an extremely simple but effective lightweight, low-cost liquid-fuel air-to-air rocket competitive with conventional solid-fuel types.[1]

Advantages of liquids over solids for the application were: (1) greater specific impulse and therefore performance compared with solid-fuel rockets of the same size; and (2), absence of smoke because liquids are cleaner burning than solids. In historical perspective, the X-4 may be considered the first "packaged" (i.e. factory-loaded) liquid air-to-air rocket because the rocket motor was sealed at the factory

---

[*] Presented at the Seventeenth History Symposium of the International Academy of Astronautics, Budapest, Hungary, 1983. This is the concluding paper in the series by Frederick I. Ordway, III and Frank H. Winter on the history of Reaction Motors, Incorporated, which in 1958 became the Reaction Motors Division of the Thiokol Chemical Corporation. This paper surveys the major projects undertaken by the Reaction Motors Division (RMD) and includes: the powerplants for the Sparrow III missile, the Bullpup, the Corvus, the Condor, the X-15, the Lifting Bodies, and the Surveyor spacecraft.

[†] National Air and Space Museum, Smithsonian Institution, Washington, D.C.

once the propellants were poured into the long, coiled spring-like tube propellant tanks. One disadvantage was that the highly corrosive nitric acid was difficult to work with and store over long periods. Interestingly, X-4 missiles were among the hardware donated to the National Air and Space Museum by Thiokol as part of their RMI collection. Possibly Lovell Lawrance acquired them in 1945 during his missile investigations in Germany and were subsequently studied by RMI for packaged liquids techniques.

The surface-to-air Taifun was another wartime German "packaged" liquid-fuel rocket and the smallest one developed at Peenemünde. It also burned hypergolics. The 75.6 in. (192 cm) long, almost 4 in. (10 cm) outside diameter Taifun was made simpler in the practically valveless method of its operation. The propellants (Salbei and "Visol," the latter another hydrocarbon), were forced into the combustion chamber by the burning of a small cordite charge. When the gases built up to a certain pressure, a disc ruptured, thus driving the fuels together. Taifun's acceleration was extremely rapid (31 g), imparting a velocity of more than 1,000 ft/sec. (304 m/sec). Burning time for the 65 lb (29.5 kg) rocket (take-off weight) was 3 sec. and the altitude reached, 50,000 ft (15 km). Taifun promised to be as easily mass-produced as solid-fuel combat rockets but its development was incomplete by the war's end.[2]

In the U.S., the "packaged" liquid-fuel concept was pursued during 1949-1950 by the Navy's Bureau of Aeronautics (BuAer) which sponsored design contracts with an eye toward achieving high performance of conventional liquid systems. One development was the LAR (Liquid Aircraft Rocket), a 5 in. (12 cm) diameter air-to-air missile tested at the Navy's Naval Ordnance Test Station (NOTS), China Lake, California. The motor, NOTS model 500C, burned the hypergolic combination red-fuming nitric acid (20% $N_2O_4$) and hydrazine with ammonium thiocyanate additive to depress its freezing point. Thermal insensitivity and impact resistance to the shock of rough handling were also LAR design features.[3]

As part of these packaged liquids investigations, BuAer assigned RMI developmental work late in 1952. RMI produced a bolted, detachable 5,000 lb (22 kN) thrust/2.5 sec unit (Figure 1). The final version was to be all-welded. By 1954-1955, this work was still experimental but did lead to a further BuAer contract in August, 1956, for the first of "a new family of packaged liquid powerplants". This was the LR44-RM-2 Guardian I engine (Thiokol designation TD-174) for the Navy's Sparrow III air-to-air missile, produced by the Raytheon Co.[4]

The background to the contract is a fascinating story recently revealed by RMI-RMD pioneer Arthur Sherman. BuAer had favored packaged liquid substitution for Sparrow III's Aerojet 1.7 KS 7800 solid engine whereas the Navy's Bureau of Weapons (BuWeps) had not. Raytheon's O. B. Randle was also against the substitution. Because of the design philosophy conflict within the Navy, a design competition was held in which Sherman and other RMI officials, BuAer, and Raytheon people were present, including Randle. As prime contractor, Raytheon held the leading cards. Randle set the tone and the ground rules for the competition. He stated bluntly: "I do not want to change the system. I do not want to change the thrust. I do not want leaks. I don't like liquids. I don't want your lousy plumbing. I do not want the center of gravity to shift. I am not about to change the launcher, so if you're going to give me a new engine give me exactly what I have." The RMI men were dumbfounded, recalls Sherman. "They left the meeting in total despondency."

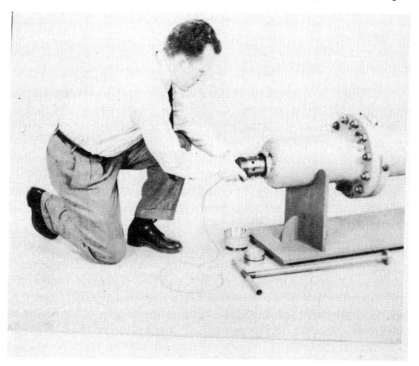

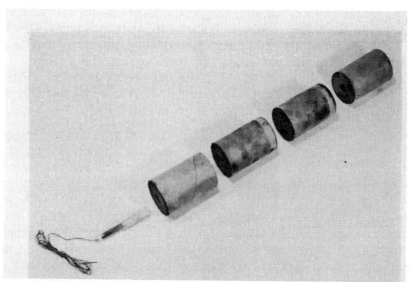

**Figure 1**  Two views of Reactions Motor's experimental packaged liquid propellant rocket producing 5000 lb (22 kN) thrust for 2.5 seconds, ca. 1952.

As a result of Randle's demands, the competition ground rules came to be changed. The new engine had to be "completely interchangeable" with the old one. Sherman, who was present in these preceedings, says that "We were thus forced to come up with a very unconventional design, based on the LAR." Sherman, who became Guardian I Project Engineer, took out a patent with others on this concept which became the basis for the Sparrow III and follow-on, packaged liquid Bullpup engines. Guardian I development commenced December 1956. Because of the "completely interchangeable" clause, according to Sherman, the thrust and duration were the same as for the solid Aerojet unit: 7,800 lb (35 kN)/1.7 sec. Guardian I's loaded weight was 133 lbs (60 kg).[5]

Apart from increased specific impulse which meant increased range, another "selling point" for the liquid over the solid motor for Sparrow was that the newer planes carrying the Sparrow went supersonic. The missiles under the wings of these planes encountered terrific heat from air friction which, according to a Navy spokesman, affected the "accuracy" of the solid-fuel Sparrows. A liquid-fuel is affected "much less by temperature," according to the spokesman. Guardian I engineer Thomas Tarbox saw the situation as far worse. Supersonic heating, he says, potentially led to propellant cracks which increased the rocket's burning area, creating sudden excess gases and explosions. "The missile was a timebomb," he warned.[6]

Sparrow III's liquid engine development was hardly trouble-free. One of the "bugs" was too much ignition delay. This was critical in that the missile was ejection-launched--it was released from the aircraft, then fired a precise moment later at a safe distance. Sparrow was also a sensitive radar-homer. Charles "Chuck" Dimmick, RMD's field representative for the Sparrow program, relates a harrowing story. The delay, he says, was not to be over two-tenths of a mil. During trials at Mount Mugu, California, Dimmick sat in the back seat of an F4H aircraft for a firsthand observation of a liquid-fuel Sparrow in action. The pilot, Marine Corps Major Daniel P. Giffens, pressed the ignition button. The ignition delay was off. Suddenly, the Sparrow surged to life but swung around to home in on its launch plane. Dimmick and the pilot had the option of ejecting immediately but resolved to stick it out. The Sparrow III found its target but luckily, its strike against the F4H was not fatal. Giffins landed as skillfully as he could at the next available airstrip. The plane's radar nose had been damaged, but otherwise the airplane was intact. Dimmick says this incident did not quite cancel the project but did set it back.[7]

Another problem was the missile's temperamental guidance system, manufactured by Raytheon. The "kick" or sudden acceleration shock interfered with it. Raytheon complained and maintained that solid-propellant motors were "a little smoother" on the take-off. Apparently RMD toned down the engine and for a brief period all looked well for the project. E. Dana Gibson was charged with setting up production. He hired veteran managers from the automobile industry very proficient in mass production. A special factory was set up at Bristol, Pennsylvania, near Thiokol corporate headquarters, according to Thomas Tarbox. He says production was so rapid and smooth that the engines were turned out faster than their shipping containers. But official Navy records show that the acceleration problem was never satisfactorily solved to accommodate the guidance and on June 30, 1960 the contract was terminated, as the missiles were "Not suitable for fleet use." By then, RMD had manufactured 400 Guardian Is. The Navy then reverted to an improved Aerojet solid unit. Nonetheless, Guardian I led to the Guardian II for the Bullpup.[8]

## BULLPUP

Bullpup was another previously solid-fuel U.S. Navy missile whose engine was substituted for an RMD packaged liquid type. However, its story was more successful than that of the Sparrow III liquid version.

Bullpup (ASM-N-7) was a short-ranged air-to-surface missile manufactured by the Martin Co., Orlando, Florida, and utilizing an Aerojet-General smokeless solid-fuel engine delivering about 25,000 lb (111 kN) thrust. The weapon was conceived during the Korean War as an inexpensive, highly accurate, non-nuclear weapon for use against small defended targets such as pillboxes, tanks, truck convoys, bridges, railroad tracks, and marshaling yards. Development took place between 1954 and 1958. The missile was also designed to be used from carrier and shore-based tactical aircraft. Initially, it was fitted with a standard 250 lb (113 kg) warhead, though later, heavier and more powerful versions were known, including a nuclear model (Air Force GAM-83B). The standard Bullpup became operational in April 1959, with the U.S. Navy's Sixth and Seventh Fleets in the Pacific and the weapon subsequently saw service in the Vietnam War. Bullpup's first guidance system was simple but effective for its short range. Bullpup was a line-of-sight weapon using a radio link. The pilot aimed the missile by a control switch or "joy stick" in the cockpit, imparting left/right and up/down directions. Tracking was facilitated by pyrotechnic flares in the boat tail.[9]

Evidently, the Navy was much taken by the advantages of RMD's packaged liquid approach from its Sparrow III experience and recognized that the adaptation could be easily made in the Bullpup. The simplicity of the Bullpup would not be compromised, nor would its safety nor reliability. The liquid offered superior performance and was capable of long-term storage in a wider latitude of temperature than the solid. The liquid system was also smokeless, like the solid Aerojet unit, and would therefore not interfere with the pilot sighting during the guidance phase. Additionally, the engine was maintenance-free, light and easy to keep in stowage, and shock-proof. It was also low-cost which was further assisted by RMD's unique restrained-firing rig in which an entire missile could be tested in a simulated free-flight, thus reducing a number of actual flight tests and holding down the cost. The missile was allowed to move freely for 18 in. (45 cm) on the rig as vibration, shock, and other data were recorded.[10]

In September, 1958, a month after the ill-fated Sparrow III's LR44 went into production, RMD received a development contract for the Guardian II or LR58 engine, for the Bullpup. The basic principles of the LR44 (Guardian I) and LR58 were identical except for size. Fourteen months later, in November 1959, LR58 reached the production stage for installation in Bullpup A. The development contract for the larger LR62 (Guardian III) for Bullpup B was let in February, 1960. Bullpup engines proved an enormous success with some 50,000 produced until the contract expired in 1967.[11]

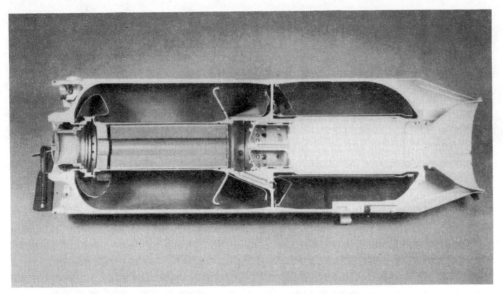

Figure 2  Cutaway of LR58 packaged liquid-propellant engine built by Reaction Motors for the Bullpup air-to-surface missile.

LR58 and LR62 burned the hypergolic inhibited red-fuming nitric acid with mixed amine (MAF). A double-base solid propellant provided the pressurizing gas which injected the fuel and oxidizer into the combustion chamber. This principle seems remarkably like that in the German surface-to-air Taifun missile of World War II. However, apart from this similar and perhaps logical engineering approach, the post-war American Guardian motors had marked differences.[12]

In the RMD motors, the liquid fuels and solid-propellant gas generator charge (a double base OGK composition) were permanently stored in high-strength hermetically sealed aluminum alloy tanks of welded construction. The mixed amine fuel (50.5% diethylenetriamine and 40.5% unsymmetrical-dimethylhydrazine and 9% acetonitrile) cooled the engines regeneratively. (The Taifun was not regeneratively-cooled though the X-4 was. More importantly, neither the Taifun nor the X-4 were constructed for long-term storage in choice of fuel or construction). Bullpup engines were additionally protected by an insulating coating of low-coat Rokide (zirconium oxide) or other ceramic sprayed over the throat and nozzle sections before welding. The only moving part in the basic engine was a piston-like initiator cartridge which moved only a fraction of an inch in order to shear off small seals, allowing the propellants to flow and activate the gas generator charge. Full thrust was achieved within one-tenth of a second upon an electrical squib triggering the dual bridge initiator. Propellant loading inlets were welded shut at the factory, insuring a storage life of five years over a wide temperature range of -80°F to 165°F

(-26°C to 73°C). Six-foot (1.8 m) drop and vibration tests were required for powerplant certification. Randomly selected samples were also subjected to 40-foot (12 m) drop tests, 30 minute bonfires; or were fired upon by 45-caliber tracer bullets to simulate battle damage, the bullets did not cause ignition and the drops and fire only caused some splitting. In any case, noted Guardian III Program Manager Alan Maier, "the liquids burn when they come in contact, but the engine does not become propulsive. And the fire can be contained with water".[13]

Table 1
BULLPUP PACKAGED LIQUID-FUEL ROCKET ENGINES

|  | LR58 | LR62 |
|---|---|---|
| Length | 40.47in (102.7 cm) | 61.20in (155.4 cm) |
| Diameter | 12.10in (30.7 cm) | 17.32in (43.9 cm) |
| Weight, loaded | 203lb (92 kg) | 563 lb (255.3 kg) |
| Weight, dry | 92lb (41.7 kg) | 205lb (92.9 kg) |
| Thrust | 12,000lb (52,800 newtons) | 30,000lb (132,000 newtons) |
| Duration | 1.9 sec | 2.3 sec |
| Impulse | 22,800lb-sec (101 kN-sec) | 69,000lb-sec (307 kN-sec) |
| Maximum safe storage Temperature | -80 to 160°F (-62°C to 71°C) | -80 to 165°F (-62°C to 73°C) |
| Maximum storage life | 5 years | 5 years |
| Altitude storage limit during storage | none | none |

This rigorous standard of manufacture and quality control contributed toward a remarkable reliability record. Out of a total of 2,695 test firings (flight and ground tests) of the LR58 alone, between 1961 and 1964, there were only three malfunctions. Reliability of both the LR58 and LR62 was rated at 0.9972%. This performance assured commercial success of the engines, though changing military needs and cutbacks eventually caused the engines to be phased out.[14]

Nonetheless, the Navy, Air Force, and other countries acquired liquid-fuel models of the Bullpup which had gone under a variety of code names that has created some confusion. Basically, the LR58 (full designation LR58-RM-2) was the powerplant for the Bullpup A (AGM-12B), which was also adopted by the Royal Navy and forces of Denmark, Norway and Turkey. The LR58 also powered the Nuclear Bullpup (AGM-12D). The U.S. Navy designation for Bullpup A was ASM-N-7a. The LR62 (LR62-RM-2) powered the Bullpup B (AGM-12C). The Air Force designation of the original Bullpup A was GAM-83A, also called White Lance to avoid using a Navy name. The Nuclear Bullpup also had the Air Force code GAM-83B. In addition to the Guardian I and II engines, there was also an experimental 50,000 lb (222 kN) thrust/5.8 dec. packaged liquid fired in September, 1958, but it was never adapted to a missile. Characteristics of the LR58 and LR62 are found in Table 1.[15]

An interesting postscript of the Bullpup story is that, by late 1975, there was a critical shortage of unsymmetrical dimethylhydrazine (UDMH) for use in the Titan 3 launch vehicle. As Bullpup motors contained about 90 lb (40 kg) each of MAF fuel which could be distilled to supply about 30 lb (13 kg) of UDMH, about 2-3,000 stored Bullpups were literally tapped by drilling holes into the sealed motors. Therefore, in this curious way, the then obsolete RMD Bullpup powerplants contributed to "high-priority space projects".[16]

## CORVUS AND CONDOR

RMI-RMD's Corvus and Condor involvements were brief. In 1957, the Navy Awarded Temco Aircraft Corporation of Dallas, Texas, the contract for the long-range Corvus XASM-N-8 air-to-surface which was to fly from Navy to Marine Corps carrier-borne aircraft against enemy surface ships or major tactical land targets 75-100 miles (120-160 km) distance. Temco chose RMI (soon to become RMD) as the engine developer, though initial versions of Corvus were apparently fitted with solid motors for aerodynamic and structural tests. For long-term storage aboard ships, RMI designed a rugged, lightweight, storable but packaged liquid-propellant system called XLR-48-RM-2, or Patriot. The propellants were completely storable, high-impulse inhibited red fuming nitric acid (IRFNA) and mixed amine fuel (MAF-1), a hypergolic combination. A small turbopump driven by a self-fed gas generator also using IRFNA and MAF-1 forced in the propellants to the regeneratively-cooled chamber. The pump was activated by a solid-propellant (HDAH) starting cartridge. Patriot's thrust was 1,030 lb (458 N) for 177 seconds (three minutes) over an extremely stable burning curve suitable for Corvus' cruising mission. Overall length of this compact engine was 38 in. (96.5 cm), maximum diameter 8.7 in. (22.2 cm), while the principal diameter (chamber) was 5.4 in. (13.7 cm). The overall dry weight was 38.2 lb (17.3 kg). James W. Fitzgerald, who instituted the tradition of assigning names as well as numerical designations to RMI engines, was the Corvus powerplant program manager. A test missile was successfully fired from a Douglas A4D at the Pacific Missile Range on July 18, 1959. Nonetheless, exactly one year later, when $80 million had been expended on the project, Corvus was canceled for then secret reasons. Afterwards, it was disclosed that Corvus was more limited in scope than newer, upcoming Navy missiles. These missiles are not identified.[17]

Condor, officially designated AGM-53A, was a much more extensive program for RMD than Corvus, and was let as another multi-million dollar contract. This was early in 1966. According to M. E. Parker, it "used up many in-house funds," particularly as it involved an advanced packaged liquid engine. Using the super exotic chlorine trifluoride (CTF) and mixed amine. The missile itself was a 2,500 lb (1,134 kg), 20 ft (6 m) long air-to-air tactical missile, 25-30 in. (63.5-76.2 cm) in diameter, with a non-nuclear warhead. It was made by North American Aviation, Inc., Columbus, Ohio, Division. Apart from the advanced engine, the Condor possessed a highly sophisticated "seeing eye" television guidance system with a high-frequency secure data-link which provided the pilot with selective target designation, precise aim-point, and positive verification of target hit. Very likely, the Navy had the advanced Condor in mind when Corvus was canceled, especially as Condor was also an air-to-surface weapon with a design range of about 100 miles (160 km).[18]

The Condor engine was a step-thrust type for high-speed take-off and slow cruise toward the target. Edward Govignon, with RMD, says its rating was 3-4,000 lb (13-18 kN) to 200 lb (0.9 kN). Condor's propulsion posed major, seemingly insoluble technical hurdles for all contractors concerned. Recalled Edward Seymour: "The corrosive properties of chloride trifluoride made inhibited red-fuming nitric acid look like a household cleanser, requiring complete shifts, from aluminum to stainless steels, and interesting phenomena like "cavitation corrosion". This particular problem was further complicated in that the propellant expulsion system was of collapsible metal bladders so that these had to be of stainless steel as well. Further, the bladders and injectors did not work as expected. "Progress was slower than planned," Seymour continues, "and it soon became apparent that development cost would exceed our estimate by a husky amount." RMD tended to push its luck too hard," said Parker, "with the result that its contract was terminated in [September] 1967".[19]

RMD was replaced by the larger propulsion firm of Rocketdyne and they too ran into the same problems. Guidance also remained troublesome. *Aviation Week* magazine for February 17, 1969 reported that the Condor's guidance and control drop tests during the summer of the previous year at the NOTS gave "disappointing results" and Rocketdyne "is not faring much better." Condor's engine, the article continues, was a two-stage system using "an extremely energetic oxidizer, believed to be chlorine trifluoride... The fuel is a hydrazine mixture. The problem is though to revolve around the need for a positive expulsion system to feed oxidizer and fuel into the engine's combustion chamber properly, regardless of maneuvering and gravitational forces acting on the missile." But despite an eventual successful test firing of a Condor from an A-6 Intruder aircraft off San Clemente, California, on February 4, 1971, the U.S. Congress soon terminated the project out of budgetary considerations, though the Navy had previously decided to initiate production.[20]

## THE X-15

The X-15 powerplant was RMI's second man-rated engine after the famous 6000C4 series. In a sense, the X-15 powerplant was a spin-off from the 6000C4. Following the success of the first X-planes, the NACA's Committee on Aerodynamics recommended in June 1952 that the agency continue these high-speed flight investigations, but at altitudes up to 50 miles (80 km) and at greater speeds, between Mach 4 and 10. From March 1954, the NACA's hypersonic wind tunnel studies at the Langley Aeronautical Laboratory soon established the new powerplant criteria. These were: a thrust of about 50,000 lb (222 kN) and propellant weight 1.5 times that of the plane. Variable thrusting over at least 50% of the thrust range and restarting were also essential. Above all, safety and reliability were paramount. In December 1955, the NACA's Research Airplane Committee selected North American Aviation, Inc. to design and build three X-15 aircraft and on February 14, 1956, RMI received a "letter of intent" from the Air Force, informing them that their proposal to build the engine had been accepted. Aerojet General, General Electric, and Bell Aircraft had also submitted proposals, but North American had reviewed these in advance and endorsed RMI's design. The contract was finalized September 7, 1956. Success with earlier X-plane powerplants weighed heavily in this decision. The RMI engine, designated XLR-99-RM-2, or Pioneer, posed far more severe requirements than the 6000C4, however. Harry W. Koch, widely experienced in aircraft, rocketry and rocket propulsion systems engineering, was named Pioneer program manager. Robert W. Seaman was the project engineer who had earlier tested the Bell X-1 powerplant.[21]

Initially, RMI/RMD made ten flight XLR-99 engines and one ground test, or "work horse" model (No. 101). Additionally, the company furnished spare parts and engineering assistance throughout the X-15 program. The first test firings were carried out at the Naval Air Rocket Test Station at Lake Denmark, New Jersey, much to the annoyance of the local residents because of the extreme noise and noise damage that resulted. The firing scheduling was also very heavy. Sometimes, as many as ten firings were made each 12-hour work day. The XLR-99 was one of the most complicated and exacting of the company's projects because it was (and remains) the largest man-rated rocket aircraft powerplant ever made. Every conceivable malfunction was artificially created during the development and testing phase. Yet, despite the time and attention lavished, when the first of the X-15s rolled out of the North American's hangar at Los Angeles International Airport on October 15, 1958, the XLR-99 was not ready. X-15 lead pilot, A. Scott Crossfield, later recalled: "The RMI XLR-99 rocket engine was steadily falling behind schedule... There were many technical locusts plaguing the RMI engineers".[22]

Koch identifies what some of these locusts were. One major problem, he says, was finding adequate cooling for the combustion chamber. "We went through Rokides [ceramic coatings] A to Z," he says, "and finally decided on the right ceramic coating. Another thing was converting from plate to a domed showerhead injector to help achieve better cooling. It all went back to cooling. The turbopump was excellent design work and the overhaul cycles were far less than what we had anticipated." As for throttling, Koch adds, "The mechanics of throttling was not a large problem, but the performance was, because we ran the entire engine at lower flow level [to find a solution to cooling]." Edward Govignon, who was assigned as a later project engineer, remembers that when he came aboard the program. "We had a lot of trouble developing the injector. A number were tried but did not work. Four or five had already been tried when I got there. The injectors had many small parts and each had to be machined to extremely close tolerances. This took time." Another major problem, noted Govignon, was that the engine shook at a certain frequency, but this observation applied to all very large rockets and was stressed because the XLR-99 was one of the largest worked on by RMI/RMD.[23]

In addition to the Lake Denmark tests, the XLR-99 was also tested at the powerplant section of the Air Force's Wright Air Development Center, Ohio, and at the Air Force System Command's Arnold Engineering Development Center, Tullahoma, Tennessee. At the latter place, 36 tests were made and completed February 16, 1960 prior to shipment of the engines to California for installation in the aircraft. The purpose of this part of the program at the Wright and Arnold Centers was for Air Force qualification and certification.[24]

Meanwhile, "By February 1958," recalled Scott Crossfield, "the XLR-99 engine was exactly one year behind schedule. . ." Countless meetings were held until L. Robert Carmen, one of the NACA's advanced design group at Edward Air Force Base, came up with a brainstorm: "I've been doing a little figuring here. Suppose that instead of waiting for the XLR-99 engine we substitute, pending its arrival, two X-1 type engines. They could be built in a few months at most." At first, there was skepticism that the old XLR-11s (Air Force designation of the 6000C4) were a viable solution. On closer examination, they were ideal. XLR-11s were not throttlable but with a total of eight chambers that could be individually ignited, a degree of throttlability was possible. The X-15 fuel tanks required no changing and dual XLR-11 engine performance was deemed sufficient for making preliminary structural, re-entry ballistic and other tests on the X-15 without major modifications. In fact, the NACA had entertained a similar solution and the plan was quickly approved.[25]

The dual Reaction Motors XLR-11s were stepped up to produce a total thrust of 16,000 lb (71 kN), about one-third the vacuum thrust of the XLR-99. The XLR-11, as used on earlier X-craft, produced 6,000 lb. or (27 kN) per engine rather than 8,000 lb, or (36 kN). X-15 aircraft #1 and #2 were fitted with XLR-11s while aircraft #3 was kept aside, as is, in readiness for the XLR-99, or "Big Engine," as it was popularly called by the X-15 engineers and pilots. (X-15 #1 was trucked to Edwards Air Force Base on October 17, 1958).[26]

The first unpowered (glide) flight, with XLR-11s, was on plane #1 on June 8, 1959 and the first powered flight with XLR-11 engines on #2 plane, took place September 17, 1959 in which pilot Scott Crossfield attained Mach 2.11 or 1,393 mph (2,241 km/hr). The highest altitude reached with XLR-11s was 136,500 ft (41,605 m), on August 12, 1960 and the fastest speed, Mach 3.4 or 2,275 mph (3,661 km/hr). That flight, made February 7, 1961 by Air Force pilot Maj. Robert M. White, was also the last X-15 flight with XLR-11s. In all, 29 powered flights were made with the twin XLR-11s (flights 2-25, 27, 29, and 31-33 (Figure 3); mission number 1 was the glide flights and flights 26, 28, and 30 were made with the XLR-99, as covered below).[27]

Phase Two of the X-15 program may be said to have begun when, on February 18, 1960, RMD announced completion of the pre-flight tests on the XLR-99. Shortly, on March 28, the first "Big Engine" XLR-99 (Serial No. 105) was sent to Edwards for installation in plane #3. However, before the first trial flights could begin, which were scheduled for the summer, the flight versions of the engine had to be put through its paces on Edwards' own Propulsion System Test Stand (PSTS) which would also familiarize the local Air Force and North American crews. In particular, they had to test the engine's throttlability and re-start capability since one of the basic requirements was that it was to be able to fire six times during a single flight--its initial firing and five restarts.[28]

**Figure 3** The X-15 with the "interim" powerplant of two Reaction Motors' XLR-11 engines, before the single chamber XLR-99 engine could be installed. (Smithsonian Institute).

There were also static tests of the engine as installed in the aircraft so that all systems could be closely monitored during a pre-flight "shakedown" test. It was during one of these static tests, on June 8, 1960, with Scott Crossfield at the cockpit of the first plane with an XLR-99 (plane #3) that an explosion occurred. Fortunately, Crossfield was not hurt though the plane was severely damaged. An intensive investigation showed that a fuel-pressure regulator failed and also its attendant relief valve. Earlier, on November 5, 1959, Crossfield had crash-landed in plane #2 when a single chamber of an XLR-11 engine failed (following the third powered flight of the X-15), so that two of three X-15's required extensive repairs at the start of the program. Nonetheless, these mishaps did not stop progress and on November 15, 1960, Crossfield successfully completed the first X-15 flight with an XLR-99. By the close of 1960, the XLR-99 engine demonstration part of the program was completed and the X-15s were turned over from the contractor, North American, to the Government (NASA and USAF). From then on, until the program ended in October 1968, on the 199th flight, the X-15 continued to exceed all previous speed and altitude records and contributed enormously toward the design of the Space Shuttle's Orbiter (Figure 4).[29]

**Figure 4**  North American X-15 rocket airplane showing the Reaction Reaction Motors' 50,000 lb (222 kN) thrust XLR-99-RM-1 engine. The man facing the camera appears to be X-15 pilot Scott Crossfield.

The XLR-99 Pioneer that made these achievements possible was a single-chamber, throttlable engine burning anhydrous ammonia and liquid oxygen fed into the combustion chamber by a turbopump driven by the decomposition products of 90 per cent hydrogen-peroxide (Figure 5). It was rated at 50,000 lb (222 kN) sea level thrust and 57,000 lb (354 kN), or 1/2 million h.p., at 45,000 ft (13,716 m). Pioneer was continuously throttlable from 50 per cent, or 24,890 lb (111 kN) minimum to 100 per cent thrust. At full throttle, with a flow rate of 13,000 lb/min (5,896 kg/min), the entire 18,000 lb (8,164 kg) fuel supply burned in 85 seconds. (This was 1,034 gallons, or 3,914 of lox and 1,445 gallons, or 5,469 of ammonia.) Some other salient characteristics of the XLR-99 are given in Table 2. The major components were: thrust chamber and injectors, gas generator, two-stage igniter, turbopump and variable governor and propellant controls, and electrical systems.[30]

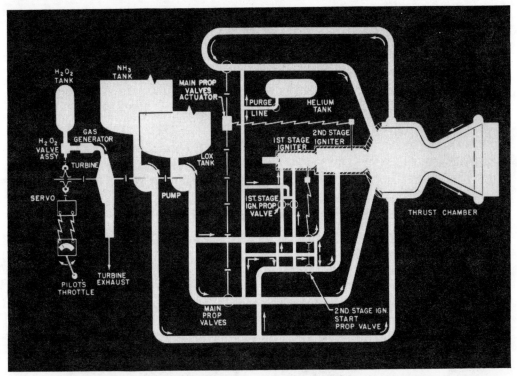

**Figure 5** Flow chart of the XLR-99 Pioneer rocket engine, produced by Reaction Motors for the X-15 rocket research airplane.

Interestingly, while Pioneer employed some striking new features in rocket engine design, it also incorporated some very old techniques worked out years before by RMI. For example, the engine used the regenerative-cooled "spaghetti" construction of the combustion chamber of contoured and welded tubes (fuel lines) which was originated by RMI's Edward A. Neu, Jr. in 1947 and developed during the period. Pioneer's fuel combination was also an early RMI innovation. It is mentioned in an RMI memo from A. Kalitinsky dated June 17, 1946 and the late Dr. Paul F. Winternitz, the company's indefatigable research chemist also worked it out during that time, as did the Navy's leading rocket expert Robert C. Truax. Experimental units using the fuel had been fired by the Jet Propulsion Laboratory in California, but, writes rocket propulsion historian John D. Clark, "RMI really worked it out in the early 50s". RMI appears to have first applied the ammonia-liquid oxygen fuel combination to their experimental XLR22-RM-2, dating from 1949-1951, and also incorporating the "spaghetti" construction. The primary object in constructing this 5,000 lb (22 kN) thrust engine, however, was to test the turbopump powered by bleeding off from the combustion chamber. The latter feature also made its way into the XLR-99. The XLR22-RM-2 may thus be said to have been the real progenitor of the X-15 powerplant, though RMI's 50,000 lb (222 kN) thrust XLR30-RM-2, or "Super Viking" engine, is usually afforded this distinction.[31]

### Table 2
### XLR-99 PIONEER ROCKET ENGINE CHARACTERISTICS

| | |
|---|---|
| Length, overall | 82in (208 cm) |
| Diameter | 39.3in (100 cm) |
| Weight, dry | 910lb (412 kg) |
| Weight, wet | 1,025lb (465 kg) |
| Thrust (overall at infinite altitude) | 58,700lb (261,097 newtons) |
| Duration | 180 sec. for max thrust |
| O/F ratio | 1.25:1 |
| Overhaul life | 1 hr (min) or 100 starts |
| Chamber pressure (full thrust) | 600 psia |
| Pump total flow rate, $O_2$ plus $HH_3$ | 198.5lb/sec (90 kg/sec) |
| Pump drive flow rate, $H_2O_2$ | 8.10lb/sec (3.6 kg/sec) |
| Turbine speed | 12,700 rpm |
| Expansion ratio | 9.8:1 |
| Nozzle design altitude | 18,500ft (5,638 m) |

The "Super Viking" certainly did play a major role toward the development of the Pioneer. It was begun in 1951 under a Navy contract as an uprated powerplant for the Viking sounding rocket. The standard Viking engine (XLR10-RM-2) produced about 20,000 lb (89 kN) thrust and burned lox and alcohol. The "Super" version delivered 50,000 lb (222 kN) thrust and burned lox and ammonia. (In 1951 a secret plan was developed to convert the peaceful, high-altitude research Viking into a 150-mile (240 km) range guided missile, but by introducing a guidance and control system rather than installing an uprated engine; however, a year later, with progress underway on the "Super Viking" engine, the plan was expanded with the aim of producing a 600-mile (965 km) range guided missile with a 1,500 lb (680 kg) warhead.) An RTV-N-12a (Research Test Vehicle) airframe for accommodating the larger, lox-ammonia engine was also designed for higher altitude or greater payload work and the new engine was expected to be completed in 1954. Turbopumps were built and tested and thrust chambers fabricated but technical problems arose and RMI could not afford to spend the hugh sums required for completing the project, so it was dropped about 1953. The Viking program itself was finished in 1955 and X-15 program initiated in the same year with the engine contract let the following year. The resumption, or follow-on of a 50,000 lb (222 kN) thrust engine based on the "Super Viking," seemed like a natural step in terms of economy, especially since the Super Viking's thrust was identical with the requirements for the X-15.[32]

When the XLR-99 project got under way, there was still considerable concern whether this combination was suitable for a man-rated engine. As Clark notes: "The great stability of the ammonia molecule made it a tough customer to burn, and from the beginning they [RMI engineers] were plagued with rough running and combustion instability. All sorts of additives to the fuel were tried in the hope of alleviating the condition, among them methylamine and acetylene. Twenty-two per cent of the latter gave smooth combustion, but was dangerously unstable, and the mixture wasn't used long. The combustion problems were eventually cured by improved injector design, but it was a long and noisy process".[33]

Apart from the propellant problems, XLR-99 designers also faced difficulties when throttling the engine. Heat transfer at all thrust levels were nearly constant, but at 30 per cent thrust there was not enough propellant to carry away the heat. This meant that the combustion chamber would overheat when the engine was throttled back. Here, regenerative-cooling was not entirely effective because of the decrease in fuel flow. The problem was finally solved with a unique combustion chamber geometry promoting fuel circulation.[34]

Another problem concerning fuel, especially in so big an engine, was the worry that in case of malfunction, a hugh amount of unburned fuel would accumulate in a fraction of a second and therefore pose an immediate risk of explosion. RMI/RMD pursued a rigorous design philosophy with safety of the pilot uppermost in mind. One approach was XLR-99's two-stage igniter system which insured vaporization and complete combustion of all residual propellants. Both first-and second-stage igniters were automatically purged with inert helium and nitrogen cooling gas every time the engine shut down. The system also enabled ignition to be safely started prior to separation of the X-15 from its carrier (B-52) aircraft. Thus, 90 per cent of the engine-starting functions were accomplished before the X-15 was committed to free-flight. The sequence began by starting the turbopumps, then the first igniter, until all was clear for the main igniter and main chamber operation. In short, the pilot had engine idling control.[35]

Another approach to maximizing safety was the propellant dump system in which the pilot could rapidly dump fuel should a malfunction occur. This was accomplished by helium forcing out the lox and ammonia. (Normally, the helium pressurized the tanks and purged the igniters.) Attention was also paid to the fuel tankage which was made of high-strength, thermal-resistant Iconel X completely welded by new techniques.[36]

Undoubtedly, one of the most noteworthy engineering and safety achievements in the design of the XLR-99 was the engine's self-monitoring system in which the engine shut down, or "safetied" itself in the event of a malfunction. This was not done by sensors and computers but by intelligent valve redundancy or redundant paths for components. Yet despite the danger of the XLR-99 becoming an over-complex "plumber's nightmare," the redundancy paid off as it achieved a remarkable 96 per cent reliability rate during nine years of record-breaking service.[37]

It was recognized early during its operational life that the XLR-99 was an exceptional, fail-safe man-rated engine with potential further applications other than the X-15. One of the suggested applications was G. R. Cramer's and H. A. Barton's 1960 concept of sending the X-15 itself into low Earth orbit. About the same time, the XLR-99 was proposed as a possible powerplant for the U.S. Air Force's X-20 (Dyna-Soar) manned orbital space glider, a predecessor to the Space Shuttle, and much hope was entertained by RMD that this would actually come about as more and more X-15 records were made. However, on December 10, 1963, X-20 was canceled by the Department of Defense.[38]

Marshall Space Flight Center's studies of XLR-99 design features toward the design of the Saturn-V engines has already been cited. XLR-99 program manager, Harry Koch, relates that Aerojet also showed a great interest in the technology and, by the late 1960s, acquired the rights of use through his good offices. Koch was then with the Aerospace Corp. at Ogden, Utah, but was able to arrange the sale of these rights. Aerojet also purchased XLR-99 drawings, blueprints, and possibly hardware, says Koch, and intended to apply the engine as a powerplant for the X-24B Lifting Body, "but they sat on it and the concept never materialized." Nor, to Koch's knowledge, did Aerojet use the engine for any other purpose. In fact, the "99" would have been too powerful for the Lifting Body.[39]

## LIFTING BODIES

Flight No. 33 of the X-15 on February 7, 1961, did not turn out to be the final mission of the venerable LR-11, descendant of "Black Betsy" of Bell X-aircraft and Douglas Skyrocket fame. Six years later, early in 1967, the old workhorse was pulled out of retirement for powering the HL-10 and M2-F3 manned "Lifting Bodies," developed by NASA's Office of Advanced Research and Technology for gathering aerodynamics data toward the design of future manned re-entry vehicles, particularly the Shuttle. Interestingly, the Lifting Bodies were carried aloft under the wing of a B-52 and then released at an altitude of 45,000 ft (13,715 m), in the manner the Bell X-1 was carried and flown. This was likewise done at the same locale, now called Edwards Air Force Base, though the Lifting Body powered flights were at speeds up to 1,000 mph (1,600 km/hr) rather than the Mach 1 domain of 20 years before.[40]

Lifting Body Program Manager John McTigue conceived the idea of utilizing either an Agena single-chamber engine or a pair of XLR-11s for propulsion in the normally unpowered Lifting Body craft. But while the Agena had a 16,000 lb (71 kN) thrust, the XLR-11 had 8,000 lb (36 kN), uprated, which was sufficient. McTigue contacted RMD manager Seymour who located six LR-11s right at Edwards. A contract was consummated and Seymour despatched specifications with Billy Arnold as a field service engineer. The Air Force had the responsibility for uprating and adapting the engines to the Lifting Bodies.[41]

There was sufficient confidence in the man-rated XLR-11s to install them into two additional Lifting Bodies, the X-24A and X-24B. The overall performance record was impressive, though it faltered at first. The initial attempt at powered flight in a Lifting Body was made October 23, 1968 in the HL-10 piloted by Maj. Jerauld R. Gentry. This flight was terminated shortly after drop-off from the B-52 when only one of the four-barreled XLR-11s ignited. The flight called for two barrels. Following safety procedures, Gentry jettisoned the remaining fuel and landed at the alternate landing site on Rosamond Dry Lake.[42]

Pilot John A. Manke had the distinction of making the first successful powered HL-10 flight November 13, 1968 when two XLR-11 barrels ignited for low-powered planned Lifting Body maneuvers. (Not all Lifting Body flights required the entire four chambers.) Manke reached Mach 0.87 but on May 9, 1969 achieved the HL-10s first supersonic flight of Mach 1.13 (724 mph or 1,165 km/hr), though stability and control were more essential to the Lifting Body program than speed. Nonetheless Maj. Peter C. Hoag made the fastest flight February 18, 1970, reaching Mach 1.86, while William H. Dana made the highest and final XLR-11 mission in the HL-10 on February 27, 1970 by ascending 90,223 ft (27,500 m).[43]

The HL-10 XLR-11 was then replaced with three Bell Aerosystems 500 lb (2.2 kN) thrust/70 second hydrogen peroxide engines for the powered approach and landing phase of the program. The more powerful XLR-11 was ideal for the powered transonic and supersonic flight phase with precise, *non-powered* maneuverable glide landings. (NASA was studying the operational trade-offs for powered vs. non-powered landings for the Shuttle, concentrating upon the more desirable non-powered option first.) In all, the HL-10 completed 21 successful flights with the XLR-11. This was matched by 22 XLR-11 flights with M2-F3, flown between 1970 and 1972; 18 powered flights with the X-24A from 1970-1971; and 24 powered flights with the X-24B between 1973 and 1975. William H. Dana made the last rocket-powered flight on the X-24B on September 23, 1975; the vehicle's final six flights were non-powered glide and landing checkouts. When NASA/Air Force Lifting Body program came to a close, the XLR-11 had logged in almost 30 years service.[44]

## SURVEYOR VERNIER, TD-339

In the heyday of America's space program during the mid-1960s, RMD tried to capture the vernier and other auxiliary rocket engine market since the giant, largely California-based corporations like Rocketdyne had already claimed much of the large-scale projects and enjoyed the facilities and territory to continue to do so. In July, 1964, RMD succeeded in securing the lucrative and prestigious sub-contract for the verniers of the unmanned Surveyor lunar probes, destined to become America's first spacecraft to soft-land on the Moon. The Jet Propulsion Laboratory (JPL) was the prime contractor. Hughes Aircraft Company developed and built the spacecraft. RMD's vernier was known as the TD-339. Donald Zimmet was the Surveyor vernier program manager, J. Wiseman the project engineering supervisor.[45]

**Figure 6** Installation of Reaction Motor's TD-339 vernier rocket engine on the Surveyor lunar landing probe, ca 1966.

The verniers were physically small. Each weighing but 5.9 lb (2.6 kg) dry weight (minus tankage), but were critically very important to the Surveyor program. Three of the gold-plated, precision-made TD-339s were fitted onto each Surveyor. They were attached to the "knee" of each of the Surveyor's three legs (Figure 6).

The purpose of the verniers was to provide propulsive power for midcourse correction maneuvers, attitude control before and during the main, 40-second retro-descent upon the lunar surface, and prime retro-power after the main solid-fuel Thiokol retro was jettisoned. The verniers thus had to work perfectly, even after Surveyor's 63-hour flight through the intense cold of space and exposure of the extreme heat of solar radiation. Further, the verniers were to be throttlable with multiple restart capability. The TD-339's developmental problems were considerable enough, but the project also initially suffered contractual problems, mentioned in Part 3.[46]

The basic TD-339 consisted of three main components: a bell-shaped thrust chamber and injector assembly, a dual propellant throttle valve, and a dual propellant on-off valve. An igniter was not necessary because the propellants were a space storable, hypergolic combination. The oxidizer was nitrogen tetroxide with ten per cent nitric oxide (MON-10); the fuel, monomethylhydrazine hydrate (MMH-$H_2O$) which contains 28 per cent water. Each Surveyor contained three individual pairs of oxidizer and fuel tanks for each vernier. A common pressurized gaseous helium tank forced the propellants into the chamber. The throttle and off-on valves were controlled by electrical signals as was the helium gas pressure valve. Each vernier was independently controlled and could be throttled from 30 to 104 lb of thrust (134-463 N). Other TD-339 characteristics are given in Table 3.[47]

### Table 3
### SURVEYOR VERNIER ENGINE CHARACTERISTICS

| | |
|---|---|
| Length | 9.3in (23.6 cm) |
| Throat diam. (I.D.), Max | 5.09in (12.9 cm) |
| Throat diam. | 0.54in (1.3 cm) |
| Weight, dry | 5.9lbs (2.6 kg) |
| Thrust | 30-140lbs (133-462.5 newtons) |
| Duration total | 4.8 min |
| Specific impulse, vacuum | 273-287lb-sec/lb (2.68-2.81 kN-sec/kg) |
| O/F Ratio | 1.5:1 |
| Minimum component endurance | 10,000 operating cycles |

The TD-339 was regeneratively-cooled but this technique produced its own heat transfer problems because of throttling, quick restarting requirements and low thrust with consequent low coolant flow. There was a danger of boiling and possible decomposition of the propellant. The complex heat transfer problems were solved with RMD's innovation of the "Voramic chamber" concept. Voramic referred to a combination of a vortex injector and ceramic throat insert and thrust chamber. A heat-resistant, thin molybdenum alloy nozzle extension was also included, as well as a fuel regulator that maintained constant high pressure to inhibit fuel boiling within the chamber cooling jacket. The fuel itself (monomethylhydrazine hydrazine and water) was chosen for its thermal stability so it could operate over all Surveyor's throttling modes in a space environment. And for insulation protection of the propellants from solar and other radiation, exposed surfaces of the verniers were coated with gold plate 0.0001 inch (0.00025 cm) thick and polished to a high luster.[48]

A project of such demanding exactness as the TD-339 was not without its share of developmental headaches. Among these were the detrimental effects of the original seals to various greases and oxidizer; paint blisterings; fuel injector distortion; fuel regulator clogging; leakage through seals; faulty solenoids; and nozzle cracking.[49]

Eventually, these and other problems were eliminated through a most vigorous test and qualifications program involving tens of thousands of seconds of firing experiments in flight-type vehicles and many hours in high-altitude vacuum simulators. These were conducted not only at the TD-339 manufacturing site at RMD's Denville, N.J. headquarters, but also at the Air Force Missile Development Center at Holloman Air Force Base, New Mexico, where a scaled-down T-2N-1 test vehicle was fired from a captive balloon at 1,450 ft or less (442 m) altitudes, with later simulated lunar descents upon the desert soil. The latter was part of the TD-339's "sea level test program".[50]

On June 1, 1966, the TD-339's technical and initial managerial tribulations were left literally far behind as *Surveyor 1* gently descended to the surface of the Moon, although it had been proceeded by the Soviet *Luna 9* which was the first spacecraft to soft-land on February 3 of the same year. Upon completion of the Surveyor program in 1968, a total of five Surveyors out of seven launched had succeeded, contributing immeasurably to the follow-up manned Apollo lunar program. The successes of Surveyor were in large measure attributed to RMD's TD-339 verniers which in themselves claimed milestones. These were:

- o   113 successful space firings of 20 Surveyor verniers.
- o   40 successful space firings by one Surveyor vernier.

    Reliable performance by 20 out of 21 flown Surveyor verniers. One vernier, on *Surveyor 2*, failed to fire for undetermined reasons.

- o   *Surveyor 5* landing was made with vernier system propellant supply pressure 110 psi (758 n/m$^2$) below design minimum.
- o   First rocket engines to restart on lunar surface were RMD's verniers on *Surveyor 5*
- o   First spacecraft flight from the lunar surface was made by the TD-339's on *Surveyor 6*[51].

The TD-339 Surveyor record is summarized in Table 4.[52]

Td-339's managers envisioned the Voyager, Mariner, MOL (Manned Orbiting Laboratory) and other candidate projects for their engine, but this never came to pass.[53]

Table 4

**VERNIER (SURVEYOR) TD-339 SPACE FIRING SUMMARY**

| Spacecraft | Midcourse Maneuver | | Landing Maneuver Burn Time (secs) | Restarts on lunar surface | |
|---|---|---|---|---|---|
| | No. of Vernier Starts | Total Vernier Burn Time (secs) | | Burn Time (secs) | Days after Landing |
| SC-1 | 1 | 20.7 | 165 | --- | --- |
| SC-2 | 40 | 51.7 | ---* | --- | --- |
| SC-3 | 1 | 4.2 | 216 | --- | --- |
| SC-4 | 1 | 10.5 | 41.9** | --- | --- |
| SC-5 | 8 | 98.4 | 109 | 0.5 | 3 |
| SC-6 | 1 | 10.3 | 182 | 2.5 | 8 |
| SC-7 | 1 | 11.3 | 201 | --- | --- |

*No landing attempt due to unsuccessful midcourse maneuver
**Spacecraft signal lost after 41.9 sec. Landing maneuver details unknown.
Of the 29 Surveyor verniers made by RMD, 21 are on the Moon.

## CONCLUSION

For all intents and purposes, the TD-339 Surveyor vernier was RMD's last major rocket project. The C-1, or "Common Engine," discussed in Part 3 of this series, appeared promising as a versatile, small liquid-fuel auxiliary rocket adaptable for a number of launch vehicle and spacecraft control systems, including the Saturn SIV-B stage and Apollo Command and Service Modules. However, RMD expended a lot of effort and money in the early developmental phase of the C-1 (which was also designated TD-345), but simply did not acquire the contract in a highly competitive market. Another RMD auxiliary rocket, the miniature pulse-rocket called "Starmite," which was capable of producing rapid on-off thrusts of 0.5 to 10 lb (2.2-45 N) for satellite attitude controls, similarly was not adopted by the NASA. During this same period, in the mid-to-late 1960s, the X-15 and Bullpup program ceased and the Condor contract was terminated. In short, RMD simply ran out of viable projects and was struggling to enter the auxiliary rocket field during very competitive times with the consequence that inevitable cutbacks in manpower had to be made to the point where RMD ceased to be profitable for the parent company, so that RMD finally closed its doors in 1972, but not for lack of talent, boldness, and vision.[54]

If there was a fault or trend within the company itself which led to its eventual demise, it is clearly intertwined in the company's project histories. The factor, which looms above all the rest, was in a word, RMI-RMD's geography. The company's location was an extremely poor one for rocket testing and in the end was to severely limit its growth. It may be remembered in Part 1 of this series that noise complaints (and eventually lawsuits) as a result of 6000C4 tests at Pompton Plains during the dynamic 1945-1946 period had already frustrated progress and forced movement of the test area to the Navy's facility at Dover. Little more than ten years later, in the wake of the post-war housing boom and "suburban creep," even the formerly remote Dover facility faced the identical problem during the day and night testing of the X-15's XLR-99. Indeed, the New Jersey Corporation was liable for property damage caused by testing the X-15 engines, as nearby homeowners charged that shock waves from the engine had cracked foundations, floors, walls, ceilings, chimneys and fireplaces. Thiokol paid the damages but failed to see, as it were, the handwriting on the wall, though individual RMD engineers did. Edward Govignon who worked on the XLR-99 project, for example, observed that "beyond a shadow of a doubt, we were severely restricted to testing large rockets and this contributed toward the demise the RMD." RMD managers too, like Dr. Edward Seymour, recognized that "our test location days for large engines were limited."[55]

Meanwhile, Rocketdyne, Aerojet, STL, and other rocket firms located in California or other expansive areas away from population centers, enjoyed all the room they needed. They developed larger engines and hence, attracted NASA and other contracts.

Just why RMI-RMD top management did not choose to expand their potential by moving to a more advantageous location for large-scale rocket testing is unknown. The financial requirements for such a move were probably a major reason. Psychologically, however, there may have been an intense collective attachment to the company's roots. RMI-RMD old-timers invariably speak of their company with great nostalgic pride, it being part of a pioneering family. And indeed RMI-RMD was America's first family in commercial liquid-fuel rocketry.

## REFERENCES

1. Willy Ley, *Rockets, Missiles, and Space Travel*, The Viking Press, N.Y. 1959, pp.242-243, 456.

2. Ley, *ibid*. pp.240-241, 454.

3. Marie Pfeiffer, "Twenty Years of Navy's Liquid Rockets", *Naval Aviation News*, January 12-13, 1962.

4. "A Short History of Reaction Motors Division Packaged Liquid Rocket Thrust Units", press release, Thiokol Chemical Corporation, (no date, ca. 1960), p.3.

5. Telephone interview, Arthur Sherman, by F. H. Winter, May 6, 1987.

6. Telephone interview, Thomas Tarbox, by F. H. Winter, April 13, 1987; Richard Witkin, "New Engine Used in Navy Missiles", *New York Times*, October 25, 1959.

7. Telephone interview, Charles "Chuck" Dimmick, by F. H. Winter, April 2, 1987.

8. Dimmick interview; Telephone interview, John Broderick (Raytheon Corporation) by F. H. Winter, May 13, 1987; U.S. Naval Air Systems Command, *Sparrow III Weapon System... Under Contract NO019-71-C-0439* [also titled *Master Plan - Sparrow III...*Vol. 2A Contracts], pp.8-2. 42.6.

9. Frederick I. Ordway, III and Ronald C. Wakeford, *International Missile and Spacecraft Guide*, McGraw-Hill Book Company, Inc. N.Y., 1960. p.27; Bill Gunston, *The Illustrated Encyclopedia of the World's Rockets & Missiles*, Crescent Books, N.Y., 1979 p.123 G. J. Geiger. "Pup with a Bite", *Ordnance*, November - December 1964 p.302.

10. Alan Maier, "Packaged Liquid Engines---America's Third Approach to Rocket Power", undated clipping, probably from the *Thiokol Astronaut*, Thiokol Chemical Corporation, ca. 1960, p.26; Bullpup Joins the Fleet; from *Thiokol Astronaut*, as above. p.25.

11. Maier, *ibid*, p.28; "A Short History", p.3; Dimmick interview.

12. Maier, *ibid*, p.27; Ley, *ibid*, pp.240-241.

13. "Bullpup: A Packaged Punch", *Aerospace Facts* (Thiokol Chemical Corporation), 1 May-June 1965, pp.6-7; John Judge, "Future Tactical Missiles May Draw on Bullpup Technology", *Missiles and Rockets*, 14, 32-33, April 13, 1964; Thiokol Chemical Corporation, *SD-102 Pocket Handbook - LR58-RM-4 [and] LR62-RM-2 Packaged Liquid Propellant Rocket Engines... Contract NOW 63-0096-f,* Bristol, Penn., 1963, *passim*; Maier, *ibid*, p.27.

14. Thiokol Chemical Corporation, Reaction Motors Division, *Spacecraft Attitude Control Engine Presentation PD-1-65*, Bristol, Penn., January 1965, p.6.

15. Gunston, *ibid*, p.123; Geiger, *ibid*, 302; David A. Anderton, "Bullpup Sets Marks for Production, Cost, Reliability", *Aviation Week*, 76 (86-87) May 7, 1962; Maier, *ibid*, p.28.

16. "UDMH Shortage Hampers Launches", *Aviation Week*, 101, 14,11

17. Gunston, *ibid*, p.122, Ordway, III, *ibid*, p.27; Tarbox interview; Dimmick interview; XLR-48-RM2 Data sheet provided by William Fitzgerald. "The Rocket, Reaction Motors Division, 11, 1, 9 January 1960; Mark S. Watson, Navy Drops New Missile", *Baltimore Sun*, July 19, 1960.

18. Interview, M. E. "Bud" Parker, by F. I. Ordway, III, March 26, 1982; Telephone interview, Edward Govignon, by F. H. Winter, April 17, 1987; "RMD Announces Condor Contracts", *Aerospace Facts*, 2, November-December 1966.

19. Govignon interview; Parker interview.

20. "Problems Balk Condor Plans", *Aviation Week*, 90, 18-19; February 1969; "Navy Tests Armed Missile", *New York Times*, February 6, 1971; "Letters", *Astronautics & Aeronautics*, 21, March 1983.

21. Michael Yaffee, "X-15 Engine Will Have Wide Space Use", *Aviation Week*, 70, 59; March 30, 1959; "The YLR99: Powerplant for the X-15", *Aerospace Facts*, 2, 2; May-June 1966. Wendell H. Stillwell, X-15 Research Results NASA, Washington, D.C., 1965, pp.12, 27; Charles V. Eppley, The Rocket Research Aircraft Program 1946-1962, Air Force Flight Test Center; Edwards AFB, California, Technical, Documentary Report No. 63-3, February 1963, p.19.

22. "X-15 Engine Undergoes Tests", *Missiles and Rockets*, 4, 24-25 December 15, 1958; A. Scott Crossfield with Clay Blair, Jr., "Always Another Dawn", The World Publishing Company, Cleveland and N.Y., 1960, p.292; Myron B. Gubitz, "Rocketship X-15", Julian Messner, N.Y., 1960, pp.135-136.

23. Telephone interview, Harry Koch, by F. H. Winter, April 15, 1987; Govignon interview; Richard Tregaskis, *'X-15 Diary'* E. P. Dutton & Co., N.Y., 1960, p.131.

24. Eugene M. Emme, (ed.), *Aeronautics and Astronautics 1915-60*, NASA, Washington, D.C., 1961, p.119; Tregaskis, *ibid*, pp.124, 132.

25. Crossfield, *ibid*, pp.292-295.

26. Gubitz, *ibid*, p.76; Tregaskis, *ibid*, 102, Eppley, *ibid*, p.23.

27. Thiokol Chemical Corporation, "X-15 Flight Record with Thiokol Engines", Thiokol Chemical Corporation publication No. *PL 1-64* January, 1964, pp.5-9.

28. Crossfield, *ibid*, p.399; Eppley, *ibid*, p.29; "First XLR99 Flight Engine Shipped to Edwards AFB", *The Rocket* (RMD), 11, April 4, 1960; Gubitz, *ibid*, pp.135-137, 275.

29. Eppley, *ibid*, p.27; Kenneth S. Kleinknecht. "The Rocket Research Airplanes"; in Eugene M. Emme (ed.), *The History of Rocket Technology*, Wayne State University Press, Detroit, 1964, p.208; 'X-15 Flight', *ibid*, p.5.

30. "The YLR99", *ibid*, p.2; Stillwell, *ibid*, p.27.

31. Memo. A. Kalitinsky, June 17, 1946, Lovell Lawrence, Jr. Papers, National Air and Space Museum; Part 2 of this series, *AAS History Series*, Vol. 11 (IAA History Symposia, Vol. 6), ed. M. R. Sharpe, 1991; John D. Clark, "Ignition! An Informal History of Liquid Rocket Propellants," Rutgers University Press, New Brunswick, N.J., 1972, pp.103-104; Gubitz, *ibid*, p.147.

32. Taffee, *ibid*, pp.60, 62, 64-65; Gubitz, *ibid*, p.147; John T. Mengel, et.al., "Rocket Research Report XI-A Phase Comparison Guidance System for Viking", Naval Research Laboratory, Washington, D.C., May 5, 1952, NRL Report 3982; pp.1, 21-22; Reaction Motors, Inc., "Liquid Propellant Rocket Engines Developed for the U.S. Military Services", looseleaf, ca. 1953.

33. Clark, *ibid*, p.104.

34. "X-15...Most Significant Manned Flight Vehicle Ever Built", *Thiokol Astronaut*, 1, 2, p.5; Yaffee, *ibid*, p.65.

35. Stillwell, *ibid*, p.29; "X-15...Most", *ibid*, pp.4-5; Kleinknecht, *ibid*, p.206; Gubitz, *ibid*, p.139; Yaffee, *ibid*, pp.62, 64.

36. "X-15...Most", *ibid*, p.5.

37. Stillwell, *ibid*, p.28; Gubitz, *ibid*, p.139; Yaffee, *ibid*, p.62.

38. Tarbox interview; Yaffee, *ibid*, pp.59-60.

39. Koch interview.

40. "New Job for an Old Work Horse", *'Aerospace Facts'*, 1, January-February 1967.

41. Richard P. Hallion, "The Evolution of Lifting Re-entry Technology", Air Force Flight Test Center History Office; Edwards AFB, California, 1983, p.32.

42. Hallion, *ibid*, p.38; List, anon., "Lifting Body Flights, 1963-1974", in "Lifting Body, Gen." file, National Air and Space Museum, passim.

43. List.

44. "HL-10 To Begin Shuttle Landing Simulation This Week", *Space Business Daily*, 89, May 19, 1970; Don Bane, "NASA's HL-10 Initial Tests OK", *Los Angeles Herald Examiner*, June 12, 1970; List.

45. "Surveyor On The Moon!" *Aerospace Facts*, 1. 6, May-June 1966; Thiokol Chemical Corporation, Reaction Motors Division, JPL/RMD Surveyor Design Review - TD339 Vernier Thrust Chamber Assembly-Ref: JPL Contract No. 950997 (Mod 4), Thiokol Chemical Corporation Denville, N.J. 1964, Chart II-2.

46. "The Radiamic Engine - Coolest Thing in Spacecraft Control", *Aerospace Facts*. 1, 4, May-June 1965; "Surveyor", *ibid*, p.6; "Thar's Gold in Them Thar Craters!" *Aerospace Facts*, 2, 10, November-December 1966.

47. "The Radiamic", *ibid*, 4; "Thar's", *ibid*, 10; Thiokol Chemical Corporation, Reaction Motors Division, Final Technical Report - Surveyor Vernier Thrust Chamber Assembly-Thiokol - RMD Model No. TD-339 RMD rpt. 4714-F1, Denville, N.J., April 19, 1968, p.2-1.

48. "The Radiamic", *ibid*, 4; "Surveyor", *ibid*, p.6; "Journey to the Moon", *Aerospace Facts*, 1, 6, July-August 1965.

49. Thiokol, Final, pp.3-60 to 3-67, 3-71 to 3-86, 4-12 to 4-16.

50. "Surveyor Verniers Pass Test", *Aerospace Facts*, 2, 11 March-April 1966; Thiokol, Final, pp.4-1 to 4-10.

51. Thiokol, Final, p.1-2.

52. Thiokol, Final, 1-3.

53. "Journey", *ibid* p.6.

54. The "'Common' Engine's Not So Common", *Aerospace Facts*, 1, 9, July-August 1965; "RMD Gets C-1 Program", *Aerospace Facts*, 1, 9. 15, November-December; Thiokol Chemical Corporation, Reaction Motors Division, The Thiokol C-1 Radiamic Engine, (Thiokol PL4-65): 1; "RMD Develops Attitude Control Rocket [Starmite]", undated clipping. "Thiokol Chemical Corporation" file, National Air and Space Museum.

55. Govignon interview; undated, untitled clipping. "Thiokol" file, National Air and Space Museum [on sites for X-15 engine damage]; "Rocket Test Damage Case Before Court," *Trenton Evening Times*, (Trenton, N.J.) March 20, 1962.

## NOTES

The following notes relate to the References as numbered:

5. The patent was No. 3,094,837, filed February 19, 1957 and granted June 25, 1963 to Arthur Sherman, Delacey Ferris, and Robertson Youngquist for "Rocket Motor".

2. The precise, overall chronology of the Sparrow III's Guardian I engine is still imperfectly known. The only datable reference to it in the Raytheon press release (ca. 1973) "Chronological Summary of Sparrow III Missile" is December 17, 1958 for the first flight of the liquid-fuel model.

10. For other cost-saving measures in manufacturing these powerplants, see John F. Judge, "Cost Consciousness Yields Bullpup Engine Savings". *Missiles and Rockets*, 11, 32, 34, October 22, 1962.

13. Guardian II (LR58) Program Manager was H. J. Schmidt. The LR58 was manufactured at Thiokol's Bristol, Penn. plant, the Guardian II (LR62) at Rockaway, N.J.

23. There literally was a special ceramic Rokide Z protecting the XLR-99's combustion chamber from flame erosion. It was provided by the Norton Company.

30. The XLR-99 was not the first American man-rated throttlable rocket engine. The old 6000C4 originally made for the Bell X-1 was not throttlable, but the Curtiss-Wright XLR-25 15,000 lb (6,818 kg) thrust engine for the Bell X-2 was.

32. Gubitz (Ref. 22, p. 147) says that Bill (William P.) Munger, former RMI Chief Engineer, suggested that the Super Viking engine should lay the groundwork for the powerplant that became the XLR-99. Munger however, denies this.

33. Clark's amusing anecdote about RMI's Lou Rapp criticizing the danger of the ammonia engine to a fellow airplane passenger who turned out to be X-15 pilot Scott Crossfield is anachronous. Dr. Lou Rapp a propellant chemist with RMI from 1951 to 1956, informs the author that this conversion took place much earlier. The discussed rocket plane was the X-1 not the X-15 and the passenger he spoke to was not Crossfield but may have been Chuck Yeager.

34. Probably the best article on the development of the XLR-99 engine from the Super Viking is Michael Yaffee's "X-15 Engine Will Have Wide Space Use" (Ref. 21).

38. Still other potential XLR-99 engine applications were upper stages for the Saturn, Centaur and other space vehicles. See *Aviation Week*, March 2, 1959, p.19.

46. For further accounts of the TD-339 Surveyor vernier initial managerial problems. see Clayton R. Koppes, "JPL and the American Space Program", Yale University Press; New Haven, 1982, pp.178-179 and Courtney G. Brooks, et. al., "Chariots for Apollo", NASA; Washington, D.C., 1979; pp.155-156.

55. So far as is known, RMI-RMD suffered only one fatility as a direct result of rocket work during 31 years. This was on the so-called "Super P", or "Super Performance" engine which has not been covered in this series. Super P was a throttlable, hydrogen peroxide/JP auxiliary rocket engine for the Navy's Vought F7U fighter, designed by Robertson Youngquist and submitted by him with William Davies about 1956. One Monday morning, the 5,000 lb (2,268 kg) maximum thrust engine exploded as a Vought Aircraft engineer was dissasembling a line that apparently had not been properly purged. The man was killed and two others injured. The theory is that over the weekend the peroxide leaked to the JP via a common turbine shaft.

## Chapter 13

## PAGES FROM THE HISTORY OF
## THE HUNGARIAN ASTRONAUTICAL SOCIETY*

### István György Nagy[†]

Dealing with the history of the host society of the 34th International Astronautical Federation Congress--first of all--we have to summarize some data about the activities of outstanding Hungarian scientists and engineers in the field of rocketry, jet propulsion and astronautics.

In the middle of the 19th Century the accomplishments of Lajos Martin (1817-1897) were of considerable importance in the domain of rocketry. In the second decade of our century the "aerial torpedo", a ramjet design of Albert Fonó (1881-1972) is worthy of note. Later, in 1928, preceding any other inventors, Fonó got the first patent in the world of the jet drive for airplanes. A good deal of effort was made in Hungary in the thirties to develop war-rockets. Zoltán Bay (1900-    ) and his co-workers in the Tungsram Laboratory developed special radar equipment from 1944 in order to get echo-signals from the Moon. This group of Hungarian scientists achieved success in February 1946.

The achievements of Tsiolkovsky, Goddard and Oberth became known in Hungary in the 1920s. Getting acquainted with some of their works popularized astronautics. In this activity, the Stella Astronomical Association (*Stella Csillagászati Egyesület*) and the Hungarian Society of Natural Sciences (*Magyar Természettudományi Társulat*) took a principal part.

Many years later the successor of the said associations, the Society for the Dissemination of Scientific Knowledge (*Tudományos Ismeretterjesztő Társulat*), adopted the idea of astronautics in Hungary again.

### THE HUNGARIAN ASTRONAUTICAL COMMITTEE

In the mid-1950s it became clear that the first artificial satellites were approaching. An initiative originated from the astronomical section of the Society for the Dissemination of Scientific Knowledge to set up a committee of astronautics as

---

\* Presented at the Seventeenth History Symposium of the International Academy of Astronautics, Budapest, Hungary, 1983.

† Hungarian Astronautical Society, Budapest, Hungary.

a basis for a later scientific society. In May 1956 the Hungarian Astronautical Committee, the predecessor of the present society, was instituted by a small group of physicists, astronomers, engineers, meteorologists, physicians and lawyers.[1] In those days, the propagation of astronautical knowledge was the main task of the committee. Some committee members composed the first Hungarian book about astronautics published in 1957.[2]

The orbiting of the first satellites aroused exceptional interest in Hungary. The committee had to meet increased demands for scientific lectures for the general public. Apart from the lectures, the committee members took an active part in giving information on matters concerning astronautics to the press, radio and television.

At this time, the work of the committee more and more exceeded the initial tasks. Various new members joined the committee. With the participation of several old and new committee members, permanent satellite tracking commenced in Hungary using both optical and radio observation methods.

In this early period, two late members of the committee distinguished themselves by their pioneering activity: The first radio tracking experiments in Hungary were realized by Endre Magyari (1900-1968), the first Hungarian research in the field of space medicine is linked with the name of Emil Galla (1909-1959). In 1958 the first astronautical conference in Hungary was organized by the committee.

## THE HUNGARIAN ASTRONAUTICAL SOCIETY

The enlarged tasks demanded the institution of an astronautical society to continue the work of the former committee within a broader range. In its present form, the Hungarian Astronautical Society (*Magyar Asztronautikai Egyesület*)--a scientific body affiliated with the Federation of Technical and Scientific Societies (*Műszaki és Természettudományi Egyesületek Szövetsége*)--was set up in December 1959. Academician Albert Fonó was elected president of the society. Although the prominent scientist was approaching his eighties in those days, he never regarded his position as a mere formal charge. He participated very actively in the everyday work of the society until the end of his life.

According to the program accepted in the statutory meeting, the main mission of the society is "to focus the attention of its members on those timely problems of astronautics which are acceptable to Hungarian science and which contribute to the peaceful uses of outer space."

The first years of the Hungarian Astronautical Society coincided with the speedy progress in space activities. The subsequent successes in spaceflight, the astounding new results in space research, space biology and space technology excited worldwide interest. Gradually these spheres of astronautical activity were definitely formulated in which Hungarian specialists could take an effective part. Under these conditions, the size of membership was well over a hundred in the early 1960s.

Two eminent leading members left their mark on the development of the society in its initial years. Ernő Nagy (1917-1969) held the secretary's office in the years 1959-1964, and György Érdi-Krausz (1899-1972) served as acting secretary from 1964 until his death. Ernő Nagy gained great distinction through his literary pursuits in astronautics. He delivered the first university courses in Hungary about the physics of rocket propulsion. György Érdi-Krausz concerned himself with astronautics and its associated sciences since the 1920s. He was a distinguished specialist in geodesy; his device, a nomographical system for transformation of co-ordinates (Navicard), was widely put in practice by astronomers in the course of optical satellite-tracking operations.

The Hungarian Astronautical Society suffered grievous losses on account of the death of both its president and its acting secretary in 1972. Under the presidentship of Iván Almár, the board of the society was reorganized in January 1973. At the same time, the organizational structure of the society was changed too; working committees and local sections were established. It is noteworthy that the organizational changes were demanded also by the accelerated development of astronautical activities in our country.

Owing to the recent Hungarian results in the domain of astronautics and the success of the first Hungarian spaceflight, with the participation of further specialists in space sciences, the board of the society was renewed again in January 1982. Some new working committees and local sections were formed.

Besides the center of the society in Budapest, local sections are active now in four other Hungarian towns: in Baja, Debrecen, Kecskemét and Sopron. The various working committees operate at present in the field of satellite geodesy, space biology and medicine, remote sensing, exploration of the Solar System, space technology, space law and the history of astronautics.

Commemorating the late leading personalities of the Hungarian Astronautical Society, the Federation of Technical and Scientific Societies established two memorial plaques: that for Albert Fonó and Ernő Nagy. The two Hungarian cosmonauts, Bertalan Farkas and Béla Magyari, were honored for the first time with one of the above-mentioned awards in 1980.

## THE ACTIVITIES OF THE SOCIETY

Because of its program, our society organized from the beginning different membership meetings, various public lectures, and motion picture presentations concerning astronautics. Apart from these, the society--chiefly in its recent period--has been the organizer or co-organizer of bigger special or general programs too: About forty conferences, symposia, colloquia, seminars and other scientific meetings have taken place between 1967 and 1983, some of them with the participation of foreign scientists and cosmonauts.[3]

Many lectures were conducted by the society. Various members displayed remarkable literary activity. The informatory publication of the society, *Asztronautikai Tájékoztató* has appeared since 1961. Altogether 38 numbers were published till now. We must emphasize our latest contribution to astronautical literature: a comprehensive special encyclopedia: *Űrhajózási Lexikon*, which came out in 1981.[4]

An interim commission of the Hungarian Astronautical Society has concerned itself with the problems of Hungarian space terminology between 1963 and 1968. Among other things, this commission worked out the Hungarian version of the Sänger Decimal Classification of Astronautics.[5] We compiled a special Decimal Classification of Satellite Geodesy, too.[6] A collection of astronautical abbreviations and acronyms was published in 1964.[7] Our society contributed to the organization of some astronautical exhibitions in the Hungarian Transport Museum, the Hungarian War History Museum, and the Budapest Planetarium.

In 1959, at the 10th International Astronautical Federation Congress, the Hungarian Astronautical Committee was represented and its delegate was acknowledged as an observer by the General Assembly of the International Astronautical Federation (IAF). The observer's status was retained in the following congresses. The Hungarian Astronautical Society became a voting member of the IAF at the 13th Congress in 1962.

In the past two decades, the society has taken part in the annual IAF congresses, International Academy of Astronautics (IAA) symposia and International Institute of Space Law (IISL) colloquia. More than fifty Hungarian papers were read at these programs. Several members of our society are active in the organs of the IAF, IAA, and IISL.[8] Many of our members engage in the activities of the Hungarian National Committee of COSPAR. It is worthy of note that the 23rd Plenary Meeting of COSPAR was held in Budapest in 1980.

## THE PRESENT AND THE FUTURE

For the time being the active membership of our society amounts to about 300. A considerable number of our members are engaged in doing work in one or other field of astronautics, chiefly in the framework of the Intercosmos Program.

The history of nearly three decades of the society reflects the imposing advance of astronautics. This course of events accelerates incessantly. The Hungarian Astronautical Society does everything possible to promote the increasing contribution of Hungarian science and technology to the further progress of astronautics as well as expedite the practical applications of space results in Hungary.

# REFERENCES

Almár I.: A MTESZ Központi Asztronautikai Szakosztályának története és jelenlegi muködése, *Csillagászati Évkönyv* vol. 1976, pp.109-114.

Nagy I. Gy.: A Központi Asztronautikai Szakosztály története, *Asztronautikai Tájékoztató* 31, (1976), pp. 3-19.

# NOTES

1. Cf. the secretary's report on the statutory meeting of the Hungarian Astronautical Committee by I. Almár in *Csillagok Világa* vol. 1956. pp.149-152.

2. *Az űrhajózás* by I. Almár, L. Aujeszky, E. Galla, I. Gy. Nagy, J. Sinka. Editor: A. Kutas-Péter.

3. General programs:
   Astronautical Symposium (1967, 1971, 1976, 1981),
   Conference on Some Timely Problems of Astronautics (1972).
   Special programs:
   Seminar on Physics of Ionosphere and Magnetosphere of the Earth (1972, 1973, 1974, 1975, 1976, 1977, 1978, 1979, 1980, 1981, 1982);
   Conference on Space Medicine (1973);
   International Symposium on Satellite Observations (1974);
   Symposium on Remote Sensing (1974, 1976, 1978, 1980);
   Seminar on Satellite Geodesy (1975, 1977, 1981);
   Conference on Satellite Geodesy (1982);
   Colloquium on Planetology (1977, 1978, 1982, 1983);
   Symposium on Physics of the Upper Atmosphere (1977, 1981);
   Seminar on Geodesy and Geodynamics (1978);
   International Symposium on Solar Physics (1977);
   International Conference on Cometary Exploration (1982).

4. Editorial Board: I. Almár (editor-in-chief), T. Echter, Cs. Ferencz, A. Horváth (editor), M. Ill, I. Gy. Nagy, Gy. Szentesi.

5. *Asztronautikai Tájékoztató* 13 (1966).

6. *.Ibid.* 15 (1967).

7. *.Ibid.* 9 (1964).

8. The participation of members of the Hungarian Astronautical Society in different, IAF, IAA, and IISL offices:
   I. Abonyi - IAF International Program Committee member 1978-1980;
   I. Almár - IAA corresponding member since 1980, IAF Vice-President since 1982;
   Gy. Barta - IAA corresponding member since 1973;
   A. Fonó - IAA corresponding member 1968-1972;
   Gy. Gál - IISL member of the Board since 1979;
   T. Gánti - IAF Bioastronautics Committee member since 1978;
   I. Herczeg - IISL member of the Board 1962-1978;
   J. Hideg - IAF Bioastronautics Committee member since 1980;
   Gy. Marx - IAA corresponding member 1968-1979, member since 1979, IAF Vice-President 1974-1976;
   I. Gy. Nagy - IAA History of Astronautics Committee member since 1979.

AAS 91-294

Chapter 14

## UNITED STATES SPACE CAMP
## AT THE ALABAMA SPACE AND ROCKET CENTER[*]

### Edward O. Buckbee and Lee Sentell[†]

Youngsters from any nation on Earth now have the opportunity to spend a week learning how people train to be astronauts and participating in a simulated Space Shuttle mission. It is the United States Space Camp, conducted each summer at the Space and Rocket Center in Huntsville, Alabama.

Already children from six nations other than the U.S. have participated in this most exciting educational program. They have come from South Africa, Panama, Canada, Switzerland, England and Venezuela.

And the camp has had additional inquiries from industry leaders in two other countries about bringing groups of youngsters to the camp in Alabama. However, sponsorship from a group is not necessary. A girl or boy from any country is accepted on the basis of his or her interest in space.

Each day opens new adventures as campers enter missions which lead them from the early days of rocketry to today's Space Shuttle and tomorrow's space stations. It's a week of computers, exercise, teamwork, astronaut gear, NASA tours, and a mock Shuttle mission.

The extraordinary camp begins with arrival, registration, and orientation on Sunday afternoon, followed by:

DAY ONE: The mission of **Rocketry Day** is to understand the principles of rocket structure, propulsion, launch and guidance. After involvement in rocketry concepts, campers begin assembly of individual model rockets. A guided tour of the Rocket Park, which former Astronaut John Glenn calls "the most complete in the world," identifies the roles of each rocket in the advancement of the space program. Most of the rockets were engineered and tested at the nearby NASA-Marshall Space Flight Center and the U.S. Army-Redstone Arsenal.

---

[*] Presented at the Seventeenth History Symposium of the International Academy of Astronautics, Budapest, Hungary, 1983.

[†] Alabama Space and Rocket Center, Huntsville, Alabama, U.S.A.

DAY TWO: Astronauts train for many years before their missions into space. The mission of **Astronaut Training Day** involves campers in a variety of specific activities, from packaged food and waste management systems to life support systems for living in space. Campers handle space suits, helmets and backpack life support units. A tour of the center where Skylab astronauts trained before the longest American missions in space is on the agenda.

DAY THREE: A highlight of U.S. Space Camp, **Gravity/No Gravity Day**, casts youngsters in the role of astronauts preparing for space walks and coping with the "zero gravity" of space. First, they step on scales to demonstrate what they would weight on the Moon, and then practice in the "Moon Walk Trainer" which reduces body weight to one-sixth normal, or one's weight on the Moon. Their team leader spins and tumbles in the Multi-Axis Trainer, similar to the device in which Mercury and Gemini astronauts were conditioned should their craft tumble out of control. They experience the sensation of up to 3 G's - or triple normal body weight - during launch in the Lunar Odyssey and rendezvous with an orbiting space station while aboard the Shuttle Spaceliner. A behind-the-scenes tour at NASA's neutral buoyancy tank - a million-gallon water tank where astronauts have trained - prepares campers for a trip to a nearby swimming pool and an underwater tank involving simulated weightlessness.

DAY FOUR: With the Marshall Space Flight Center in the foreground of planning the nation's first orbiting space station, campers study development of large space structures and their benefits for mankind. The mission of **Tomorrow's Technology Day** outlines a variety of careers in the aerospace field - apart from that of astronaut - which will be available in the future. Courses of study are also recommended. Activity with computers, a part of each day at camp, is intensified. Campers also launch their model rockets.

DAY FIVE: The culmination of the week is **Space Shuttle Mission Day**, an experience about which most youngsters can only dream. Each team of 10 youngsters is divided into crews for the Shuttle spacecraft and Mission Control. Using equipment acquired from NASA, team members conduct a simulated mission, beginning with checkout, countdown, launch, orbit and return to Earth. Each team's performance depends upon how well campers apply principles learned earlier in the week. After a lunch of freeze-dried food like the astronauts eat, campers in concluding ceremonies receive their U.S. Space Camp wings. In mid-afternoon, they depart for home, ending the most incredible week a youngster can experience.

The program, which begins its third year next spring, had already drawn participation of 2,181 girls and boys from 46 states and six other countries. (See list of states and nations.)

It is modeled after successful outdoor science camps that focus on non-traditional forms of teaching with innovative hands-on experience while in a traditional camp-like setting with team participation and competition.

**Figure 1** View of Space and Rocket Center and exhibits.

**Figure 2** A camper bounces high in the air using the moon-walk trainer.

**Figure 3** Campers monitor the simulation of a space mission from mission control using a full scale mockup of the space shuttle orbiter.

**Figure 4** Campers in the space shuttle orbiter mockup assuming the roles of pilot and commander in a simulated space mission.

Space Camp grew out of the desire of Dr. Wernher von Braun to interest youngsters at an early age in aerospace technology. He asked, "Why not have a science camp like others have sports camps?" That discussion with Space Museum Director Edward O. Buckbee resulted in a pilot program in 1981 that helped shape material for the first camps in 1982.

Among the notables involved in staffing the program is Konrad Dannenberg, who served as deputy director of the Saturn-V team under von Braun. He spends time each week with small groups of youngsters to explain and demonstrate the principles of rocketry. Other personnel include science teachers and exceptional college engineering students.

The following list indicates the home states and nations of the participants of Space Camp in 1982 and 1983:

| | | | |
|---|---|---|---|
| Alabama | 408 | New Hampshire | 3 |
| Alaska | 3 | New Jersey | 38 |
| Arizona | 4 | New Mexico | 10 |
| Arkansas | 56 | New York | 56 |
| California | 96 | North Carolina | 51 |
| Colorado | 20 | North Dakota | 0 |
| Connecticut | 22 | Ohio | 75 |
| Delaware | 2 | Oklahoma | 50 |
| District of Columbia | 0 | Pennsylvania | 31 |
| Florida | 116 | Rhode Island | 1 |
| Georgia | 165 | South Carolina | 39 |
| Hawaii | 0 | South Dakota | 1 |
| Idaho | 3 | Tennessee | 142 |
| Illinois | 74 | Texas | 115 |
| Indiana | 59 | Utah | 2 |
| Iowa | 10 | Vermont | 0 |
| Kansas | 29 | Virginia | 68 |
| Kentucky | 49 | Washington | 10 |
| Louisiana | 54 | West Virginia | 14 |
| Maine | 6 | Wisconsin | 21 |
| Maryland | 32 | Wyoming | 5 |
| Massachusetts | 12 | | |
| Michigan | 74 | South Africa | 3 |
| Minnesota | 12 | Panama | 1 |
| Mississippi | 54 | Canada | 14 |
| Missouri | 54 | Switzerland | 1 |
| Montana | 2 | England | 1 |
| Nebraska | 4 | Venezuela | 2 |
| Nevada | 4 | | |

| | |
|---|---|
| Total | 2181 |

## BIBLIOGRAPHY

1. "A Space Camp for Astronaut Hopefuls", *New York Times*, July 2, 1982.
2. "Space Camp: A Summer Star Trip", *Science Digest*, February 1983.
3. "Space Camp", *Odyssey*, April 1983.
4. "Suited for Space", *National Geographic World*, May 1983.
5. "Far-Outward Bound", *Chicago Tribune*, June 13, 1983.
6. "Big Summer at Space Camp", *Newsweek*, June 13, 1983.
7. "Space Camp Blasts Off", *New York Post*, June 18, 1983.
8. "The Kids at Ed Buckbee's Alabama Camp Aren't Spaced Out--They're Just High on Rocketary", *People*, August 8, 1983.

# Part V

# PIONEERS OF ROCKETRY AND ASTRONAUTICS

# Chapter 15

# A LIFE DEVOTED TO ASTRONAUTICS: DR. OLGIERD WOŁCZEK (1922-1982)*

## M. Subotowicz[†]

**BIOGRAPHICAL REMARKS**

Dr. Olgierd Wołczek died on August 24, 1982 in Warsaw. From 1971 he edited the scientific-popular Polish bi-monthly *Astronautyka* and also from 1973, the scientific journal of the Polish Astronautical Society (PAS) *Postępy Astronautyki* (Progress in Astronautics). He was one of the founders of the PAS (1954), then its General Secretary for 10 years and later the deputy of the President of PAS for years. He was very active also in the field of scientific research in astronautics and space physics. The scope and breadth of his knowledge can be seen in his 22 books and 34 papers on astronautics and space physics, and 10 books and 14 papers on nuclear physics and other subjects. He published also several hundred papers in popular journals, and took part several hundred times in radio and television programs.

The variety of subjects in which Dr. O. Wołczek was interested, the richness of his mind and his deep humanity, can be seen even in his three papers prepared for the 33rd IAF Congress. They were written on novel aspects of cometary research, on problems of extraterrestrial influences on biological evolution, and on the importance of astronautics for material, spiritual and biological development of mankind. His PhD-degree (1963) was based on his research in nuclear spectroscopy. But astronautics became the main interest and aim of his life.

In the first half of 1954 I published in the Polish scientific-popular monthly *Problemy* the article on astronautics and made the appeal that "It is time to institute the Polish Astronautical Society". The first who supported my appeal was Dr. O. Wołczek. During the meeting at the Institute of Physics of the Warsaw University in December 1954, several young physicists decided to institute the Polish Astronautical Society (PAS). Dr. O. Wołczek was among the several charter members of PAS. He was the most active member of the PAS in the organizing and popularizing scientific work. He wrote in 1976: "I was deeply interested in astronautics as a 9-

---

\* Presented at the Seventeenth History Symposium of the International Academy of Astronautics, Budapest, Hungary, 1983.

† Polish Astronautical Society in Warsaw and Institute of Physics, University M. Curie-Sklodowska in Lublin, Poland.

year-old boy. The first step to participate in cosmic research was to institute PAS. Together with Professor Subotowicz we initiated PAS three years before the first Sputnik."

**Figure 1** Dr. O. Wołczek in his office.

When doing the work of the main specialist in the office of the government's plenipotentiary for nuclear energy, Dr. O. Wołczek wrote that he was looking "for the possibility to be engaged completely in space research." All his scientific activity in nuclear physics was taken "as the preparation for space research."

In the several last years Dr. O. Wołczek worked in the Air Medicine Institute. There he could spend all his time working on space research. In space activity he found the sense and joy of his life. One could see this in his scientific work, in his enormous knowledge on different problems in astronautics, in his engagement in social work connected with PAS, in his joy in popularizing astronautics in his books, in radio and television talks, in hundreds of the popular articles he published in the PAS bimonthly *Astronautyka*, and in his innumerable popular lectures.

He received international recognition while sponsoring very actively international cooperation in astronautics. Dr. O. Wołczek was a corresponding member of the International Academy of Astronautics (IAA) in Paris, and as a member of several committees of the International Astronautical Federation (IAF) and IAA, he took part in more than 20 IAF Congresses as the author of interesting scientific papers. He was an honorary member of several foreign astronautical societies and very well known among the people dealing with astronautics. His books were translated into English, Russian, French, Hungarian, German and Japanese languages. He was invited to present lectures in Vienna, Berlin, Dresden, South Dakota (U.S.A.), Winnipeg (Canada), Kaluga (U.S.S.R.) and Munich (F.G.R.).

Dr. O. Wołczek's best books, in his opinion, were: *Isotopes and Men's Duty* (1955), *Secrets Taken from the Sky* (1962), *Interplanetary Flight* (1973 and 1980), *Birth and Development of the Solar System* (1979), *M. Skłodowska-Curie* (1975), and *Cosmic Scenes of life* (1982).

For seven years he was the chief editor of the *Proceedings of the IAF Congresses* published in cooperation with Gauthier-Villars, Dunod and Pergamon Press by Polish Scientific Publishers in Warsaw.

His scientific production is contained in 15 papers on nuclear spectroscopy and 34 papers on space physics, dealing with planetology, evolution of matter in the Solar System, nuclear propulsion rockets, influence of cosmic factors in the origin and evolution of living organisms, as well as the impact of astronautics on the further development of science, engineering, and civilization.

Dr. Olgierd Wołczek, born on April 3, 1922 in Toruń (Thorn), started his education in Katowice in Silesia. He took part in the second World War in September 1939. After being arrested and then released by the Gestapo, he spent all the occupation in Czestochowa, where he finished the secret secondary school. From 1945 until 1949 he studied chemistry at the Technical University in Łódź. In his diploma work he dealt with the problems of the separation of uranium isotopes. From 1949 until 1955 he was an assistant at the Warsaw University, and from 1955 until 1968 he worked in the Institute of Nuclear Research. From 1968 until 1976 he was engaged as the main specialist in the office of the governments plenipotentiary for nuclear energy. From 1976 until his death he worked at the Air Medicine Institute in Warsaw.

He was very laborious and all his passion in work he devoted to astronautics. During his scientific work in experimental nuclear physics he lost one eye about 30 years ago, but this could not stop his activity in astronautics. Being engaged in nuclear physics he devoted his knowledge to astronautics. He dropped his job at the Nuclear Physics Institute, and, stimulated by his interest, started to work in astronautics and space physics. He was the only author of his books and papers; none were written as a common paper or book with another author. His knowledge of very different subjects in astronautics was enormous. He was very familiar with many important topics in space physics, astronomy and even space biology. In some subjects, e.g., planetology, he was the best specialist in Poland.

Dealing with almost all astronautics and space physics on a popular level, his scientific activity of a qualitative character can be placed in the following four subjects:

1. Nuclear energy in rocketry,
2. Impact of astronautics; various non-selected problems in astronautics,
3. Evolution of matter; planetology,
4. Life in the Universe.

During his several last years Dr. Wołczek dealt mainly with subjects (3) and (4) mentioned above. We shall review the scientific papers of Dr. O. Wołczek according to this list of subjects.

Independent of his engagement in the physical sciences, I would like to stress the role of humanism in Dr. Wołczek's creations and his personality. His knowledge in science, physics, chemistry and astronomy was very impressive. But his erudition in classical education, in literature, arts, and music was also outstanding. He liked mountaineering. He enjoyed traveling as the means to become acquainted with other people and their work. He knew activrecords several European languages (English, Russian, German and French) and made himself understood in Italian and Spanish. He saw in the social and philosophical consequences of astronautics the synthesis of two important trends in the development of mankind: the natural, or scientific, and the humanistic ones. He represented the contradiction in the narrow specialization in science. He was thinking in human affairs on the scale of the planet or on even a greater scale. Thanks to astronautics it will be possible to conquer the Solar System, to realize the centuries-old dream of the space visionaries. Dr. Wołczek believed that this would be the first step only. The following steps will be the penetration of mankind into the Galaxy, traveling to the stars. It will be the new stage in the evolution of the human species, the stage of the *homo galacticus*. Astronautics is the prerequisite of the development of the human civilization and culture. Dr. Wołczek wrote: "The prolonged stay of the human being in space will deeply influence not only his mind but also his body. . . . The intense influence of deep space must induce extraordinary bodily, mental, and spiritual changes in human beings. If man will withstand the ruthless pressure of the galactic conditions and defy the infinity of space, he will emerge totally altered - as a new man, *homo galacticus*."

All who met and were stimulated by this dynamic man, Olgierd Wołczek, mourn his untimely death in his 60th year. Astronautics was for him not only a way of life, but its greatest adventure.

## SCIENTIFIC ACTIVITY

### Nuclear Energy in Rocketry

In two papers Dr. Wołczek proposed to build a cold atomic reactor instead of a thermonuclear one and to use it in a nuclear rocket.[1,2] In the thermonuclear reactor, the Coulomb barrier between the hydrogen isotopes will be overcome thermally at temperatures higher than $1.5 \times 10^7$ K. In the case of catalytic synthesis, the nuclei can be subjected to synthesis through mu-mesons. The radius of the mu-meso-atom is 20 times smaller than that of the ordinary hydrogen atom. These meso-atoms can be close enough to realize mutual synthesis in low-temperature meso-protonium, -deuterium, and -tritium. This project requires large sources of mu-mesons, produced in linear accelerators. The author discussed the shape and performance of the fusion reactor in the space rocket. Its main disadvantage is the short life-time of the meso-atoms.

In one paper, Dr. O. Wołczek dealt with the problem of the exploitation of nuclear energy for experimental purposes in cosmic space.[3] Artificial sources of this energy--explosive and non-explosive--could serve studies concerning cosmic space properties by determining the matter present in it and by permitting the exploration of the properties of this matter. They permitted also definite exploration of meteoric swarms. Dr. Wołczek proposed to form a special cosmic laboratory that would perform work on experimental astrophysics and star models with respect to artificial sources of nuclear energy. The laboratory would begin with the practical exploitation of nuclear energy for the generation of electric energy on the Earth's surface, and then extending it to the problem of so-called anti-matter and its properties.

### Various Problems, Impact of Astronautics

The main interest of Dr. O. Wołczek in astronautics was connected with (a) application of nuclear physics achievements in rocketry, (b) planetology and physical evolution of matter, and (c) life in the Universe. This does not mean that other topics of astronautics were outside his interest. Let us present his several achievements in dealing with the free radicals in rocket propulsion, with technical realization of subgravity and weightlessness, with the methods of measurement of the distances of space rockets, and with the impact of astronautics on science, technology, human civilization and culture.

In another paper, the possibilities of using nuclear and corpuscular non-nuclear radiation for producing free radicals were examined.[4] "Free radicals" mean atoms or groups of atoms in the metastable state. The reactions between free radicals are accompanied by considerable release of energy. The amount of energy is often one range of magnitude greater than the corresponding amounts of energy released in the most advantageous chemical reactions, e.g.: $H_2 + 1/2\ O_2 \rightarrow H_2O$ + energy released 3,810 cal/g (chemical reaction) and $2H \rightarrow H_2$ + energy released 51,210 cal/g (reaction of free radicals). The author considered also the use of beta and alpha rays from isotopes and radiation produced inside a nuclear fission reactor to produce free radicals. The proposed method would seem to be of some practical value.

Two papers considered the technical realization of subgravity and weightlessness on Earth and under full effects of the force of gravity.[5,6] He presented a comparatively simple method using centrifuges for devices operating on the same basis. All such devices should be stationed in a vertical position so that their axis of rotation would be parallel to the surface of the Earth. In this way the inertial (centrifugal) force of rotation is added to the gravitation force alternatively weakening or strengthening its effects. Practical methods were given for realization of rapid transitions from multi-g field to states of subgravity and weightlessness and vice versa, with the aim of conducting research work in space techniques and medicine. He proposed the construction and exploitation of certain apparatus for producing intermittent subgravity and weightlessness lasting for longer periods of time, on the order of hours and more.

In one paper, Dr. O. Wołczek proposed to use a simple television check altimeter for use in cis-lunar space.[7] This distance of the satellite circling the Earth at a height about 400 km would be measured with the precision of approximately ±160 m.

In several papers, Dr. O. Wołczek discussed important contributions of outer space research to the development of science and technology, due to the need to overcome extreme physical conditions which do not occur on the Earth. Direct access to outer space was a fact of basic importance.[8-10] Astronautics has brought about large-scale research on many phenomena connected with the preparation for and carrying out of outer space flights; fundamental technical problems have been solved by the introduction of new plastics, methods, construction, apparatus, etc. Outer space in the solar system and its heavenly bodies are subject to direct exploration.[8]

Space research required particularly vast development of science and technology which could not have been realized without it. The progress has contributed to the development of civilization, which is, however, conditioned by further development of astronautics.

Astronautics is of fundamental importance to the study of matter, which outside the Earth appears in forms other than on our planet.[9] The problem is inseparably connected with the mystery of life and possible existence of intelligent creatures other than men. There exists the need to abandon anthropomorphic and geomorphic habits. The Earth constitutes merely a trivial fragment of space, for which it cannot serve either as a pattern or a point of relation. The start of man to step over the space threshold is the epochal event in the history of mankind. The long and sustained stay of human beings in space and on other celestial bodes can, and probably will, have profound repercussions on the mentality and development of local and more general civilizations and cultures. Therefore, problems of the evolution of *homo cosmicus* and *homo galacticus* were presented in some detail and the impact of such evolution on mankind was outlined by the author. Dr. Wołczek was of the opinion that true progress, the improvement of mankind, is connected more with mental and ethical activity than with structural and physical changes of human organisms.

Spaceflight involves the cooperation of many institutions, many factories, and many people; it developed not only as a result of an international competition. It implies something mystical; man perceives infinite depth and the vastness of space rousing his imagination. It is something totally new, something extremely stimulating. Humanity cannot develop further without access to space. Penetrating into the cosmos, man will look from the most wide perspective at matter, at the whole Universe, at the ultimate fate of mankind. Involved in this exceptional situation he cannot remain internally unchanged.

## Evolution of Matter, Planetology

The problem of matter in the Universe, its evolution and shape in various conditions was the main scientific interest of Dr. Wołczek during at least two decades. The new possibilities of the investigation of matter allows deep insight into its different shapes and forms in extraterrestrial conditions, sometimes in extreme conditions. Four principal interactions determine all organization of matter: gravitational, electromagnetic, weak and strong interactions; their relative strengths are: 10 : 10 : 10 : 1. Weak interactions are responsible for the radioactive decay of nuclei and that of elementary particles. Strong interactions determine the existence of the nuclei and their interactions with mesons and baryons. The electromagnetic interactions determine the processes in and with the electronic shells of atoms and molecules. It means that these interactions determine the properties of gases, plasma, minerals and living matter. New information on matter, its distribution and its dynamics, is delivered now by astronautical means: space, interplanetary, Moon, planetary and galactic probes. We understand better the inertial structure of the Earth and its atmosphere. In one paper, Dr. Wołczek described results of the investigations of the Moon and planets, and discussed the cosmogonical implications of this research.[11] It is important to elaborate the perfect methods of the remote, automatic chemical analysis of matter on various planets and in interplanetary space by using specialized probes.[11-13]

In other papers, Dr. Wołczek dealt with the problems of evolution of the solar system.[15,16] He discussed the size of the Solar System and forces responsible for its evolution, the contemporary and primary distribution of matter. There are some peculiarities of the process of organization of matter in the primary circumsolar nebula. The aggregates were formed from dispersed matter. The role of the solar wind and magnetic field in the early Solar System was described, as well as the time of formation of planets and coupling between planetary masses and orbits, their longevity and that of the whole Solar System.

One paper discussed various theories of the origin of Solar System.[16] Several proposed extraterrestrial experiments that could serve as the models of the processes in the interior of the stars, e.g., thermonuclear energy production. There would be necessary the energy supply, its concentrations (lasers), plasma generators, diagnostic and measuring apparatus. The main energy source would be the Sun. Plasma could be contained by magnetic and electric fields. Principally it would be possible to realize experiments in very high temperatures and densities of matter.[16]

Two papers discussed the highlights of the evolution of exploration of matter of the Solar System by space techniques.[17-18] The morphologic and topographic investigations of the planets enable better recognition:

1. Of the structure and dynamics of the planets' surface and its connection with history of this and other planets,
2. Of the influence of the atmosphere, cosmic environments and Sun on this structure, and dynamics of the climate in the past, now and in the future.

It discussed the role of cosmo-chronological measurements, the search for life, and underground soil investigations. The up-to-date (1975) probes have many disadvantages (reliability, electronics, energy supply, radioisotope thermoelectric generators, energy conversion rate).

Especially important was the investigation of terrestrial planets and their surface morphology.[19] Dr. Wołczek discussed other research goals, such as investigations of surface layers and factors which were and are shaping the upper strata of planetary crusts. Subsequently Dr. Wołczek stressed the very important role of simulators.[19] In connection with Mars exploration, emphasis on dust interaction with landers and their instrumentation was mentioned.

The morphology of the Martian surface indicated that in the distant past a denser atmosphere containing appreciable amounts of water existed on Mars. Evidence was found that in the period earlier than 600 million years ago the rate of surface erosion was very great. On the surface of Mars, various channels exist which could be formed by the action of fluid water. In one paper, Dr. Wołczek expressed the hypothesis that although irreversible decomposition of great quantities of water could take place in the past, considerable amounts of it may persist now trapped in the form of permafrost and in polar caps of Mars.[20] There are some indications, of the existence on Mars, of surface transient atmospheric precipitations and rivers, influenced by internal (volcanic and tectonic activity) and cosmic factors (changes of Mars rotational axis inclination, the Sun's luminosity fluctuations, penetration of the Solar System into condensations of interstellar matter).

Consequently, developing his knowledge in planetology, Dr. Wołczek discussed the contemporary view of terrestrial planets.[21] Their overall characteristics, including the Moon, testify that all these bodies constitute a very distinct and well defined group from the very beginning of their formation: in a relatively limited space and in the vicinity of the central star of the Solar System. Due to their relatively small mass they could not retain hydrogen and helium, constituting the bulk of the system's matter. These planets were formed from materials of relatively high condensation temperature, mainly iron and silicon compounds. The extent of the interior planetary activity on the surface and in the atmosphere depended strongly on the size and mass of these celestial bodies. Solar radiation, chemical composition, magnetic and electric fields, internal activity and the presence or lack of life shaped the evolution of the atmospheres and the surfaces of these planets. Their evolution was affected by the surrounding present and past medium, the central star, the whole Solar System and the rest of the Universe. All these phenomena could influence also the evolution of life. It is a typical feature of Dr. Wołczek's treatment that he analyzed the phenomena in their mutual interaction and dependence, even on the scale of the whole Universe.[21]

Recent investigations of extraterrestrial matter were performed outside the Earth using almost exclusively passive means with some exceptions (meteorites, samples of lunar material). Such means delivered much information on the chemical constitution of extraterrestrial matter, physical state and structure (mineral, petrological, etc.), field characteristics (magnetic, electrical, gravitational), radiation, structure and dynamics on the surface and interior of celestial bodies, and possible occurrence of life. But we have no data on the evolution of foreign matter, on dense and opaque tropospheres of many planets, on boundary layers between atmospheres and liquid parts of Jupiter and Saturn. Dr. Wołczek proposed the use of active means in these investigations, such as: sources of intense coherent radiation, entry and impacting pellets, penetration and impact probes, aerostatic and aerodynamic devices, electrical discharges, chemical and nuclear explosions, enabling acquisition of unambiguous information directly at the site being the aim of the investigations.[22]

Dr. Wołczek described penetrators and other terradynamic devices to investigate effectively the upper layers of the planetary crusts (Mars, Venus, Moon).[23] He presented the possibility of investigation of physico-mechanical, thermal, magnetic and electric properties and of examination of composition, structure and dynamics of the subsurface crust layer by using the terradynamic devices, especially on Venus.

The present knowledge of the Jovian planets and their natural satellites is far from satisfactory. Dr. Wołczek proposed complex single-planet missions and stationary planetary orbiters, equipped with advanced instrumentation and with many penetrating devices enabling the execution of numerous investigations in the planetary atmospheres and on the surface of the planet.[24] For investigation of atmospheres, free-falling probes and self-propelling probes were recommended. Grenades would enable investigations of deeper atmospheric layers. Investigations of Jovian planets' moons may be executed by means of soft-landers. Extensive experimental and computational work should precede the practical realization of the proposed Jovian planetary probes.[24]

In one of his papers, Dr. Wołczek suggested a mission to the asteroids.[25] Their direct investigation with the aid of astronautical means may contribute considerably to the solution of several fundamental problems concerning the Solar System and its evolution. Probably unmanned exploration of the asteroid belt will be feasible in this century. A specially equipped probe for this mission was proposed.[25]

Dr. Wołczek presented planets as dynamic systems proposing more precise definition of planets, where the criterion of mass is supplemented by dynamic attributes of the celestial bodies.[26-28] The author mentioned three dynamic systems: internal, planetary or *global*, *atmospheric* and *magnetospheric*. The fourth one is the *biospheric* system. These four systems create one dynamic all-planetary system. There exists the hierarchy of these four dynamic systems, a very close feed-back and interdependence among them leading to the creation of a superior dynamic planet-wide system. The author discussed the birth and evolution of these dynamic systems on terrestrial planets.[26] These systems develop under the influence of internal and cosmic factors.

The role of boundary layers between the above systems was discussed.[27] Boundary layers never behave as passive interphase division but play an active role in the conservation of the separate character of the above systems and in their evolution. The boundary layers function as filters between planetary systems, as zones of coupling of different phases and sometimes as barriers. They protect media and systems enabling their dynamic stabilization. On the Earth the magnetospheric, ozonospheric and lithospheric barriers are the most important. There exists some hierarchy between the barriers. The existence of barriers was a precondition of the rise and evolution of the biospheric system.[27]

Dr. Wołczek called our attention to the fact that dynamic planetary systems are not closed, but open--undergoing continuous and fluctuating influences from the outside. The phases formed in these systems are of the dissipative type.[27] The formation and evolution of these phases in the Solar System from its beginning were discussed. The biosphere has particular character with its self-organization and evolution processes. In Dr. Wołczek's opinion there was some advantage in analyzing the past and future of the Solar System, based on general assumptions about the dynamic systems in our planetary system.[29]

In connection with the above idea of dynamic systems, new research methods and devices were proposed which will enable the creation of complex and more truthful pictures of planets as dynamic systems.[30-31] Dr. Wołczek emphasized the extreme scientific value of direct investigation of foreign planetary systems with the aid of interstellar probes.[30-31] He discussed the nature of extra-solar planets and methods of detection of extra-solar planetary systems. The earliest interstellar flight will occur in about 30-50 years from now.

Dr. Wołczek's last paper dealt with planetology and novel aspects of cometary research, presented during the 33rd IAF Congress in Paris.[32] The origin of comets is far from being elucidated. The complex internal dynamics of active comets awaits a detailed explanation. What is the nature of the residual material remaining after the outgasing of the cometary nucleus? There is organic matter that could play a particular role in the origin and development of life on celestial bodies. It would be very important to deliver to the Earth the material from different parts of a comet. The investigation of this material will yield very valuable information pertaining to a biogenic synthesis of biologically important organic compounds and the origin of life.

Some meteoroid streams can be identified with comet remnants or with debris lost by comets along their paths. Studying cometary material is connected with the investigation of selected meteoroids. But Dr. Wołczek believed capture of larger pieces of cometary material would be of principal importance.

## Life in the Universe

The problem of life in the Universe was, together with the planetology, the main topic of Dr. Wołczek's interest. We present now the main ideas of his several papers dealing with the problems of life. On the search for life on Mars, Dr. Wołczek pointed out that analysis of the appearance of life and its development outside of the Earth is conducted by investigating the conditions associated with the birth of life and its evolution on the Earth's surface.[33] Then the possibilities of the survival of life in extreme conditions is discussed, also in the Martian environment. The search for organic compounds on Mars and for metabolism proved unsuccessful, but even this negative result may help us to understand better the history of life development and evolution on the Earth.

Dr. Wołczek considered life as a global phenomenon, namely, the occurrence of life and rise of intelligent beings were stages of evolution of cosmic matter.[34] The terrestrial living system is based on carbon and water. Life might originate after passing by the Earth through many specific stages during very long lasting processes and very specific conditions: presence of a primitive reducing atmosphere, evolution of photosynthesizing organisms and an atmosphere containing rising quantities of oxygen, occurrence of continental drift, appearance of magnetodynamic shock waves and ozonosphere, magnetic field reversal and existence of various cosmic interactions. These conditions and facts exerted a deep influence on the course of this process. Despite essential links with the whole, man distinguishes himself fundamentally from all other living organisms by his exceptional evolutionary abilities. Due to the development of life and of intelligent beings the Earth has gained a special position in the Solar System, but most probably it is not an exception in the whole Universe. The quality of planets orbiting around other stars, where living organisms and even intelligent beings may have appeared, is not insignificant. The conquest of space has opened to mankind new and maybe exceptionally promising perspectives of further development.

Dr. Wołczek defined the properties of living organisms:[35]

1. Conservation of the individuality and self-limitation,
2. Exchange of matter, energy and information with the environment and their optimization,
3. Reproduction and transfer of the genetic information,
4. Evolution.

We do not know the nature of life, but probably it belongs to one of the fundamental properties of matter. It appears in suitable conditions in a similar way as the organic compounds.

The left-hand chirality of the molecules essential for the existence and evolution of living organisms is one of their fundamental features.[36] According to the opinion of the author the discrimination in favor of the left-hand chirality of the biogenic organic compounds reflects the deviation from symmetry (left-right hand symmetry) typical also of the non-living matter. It is a fundamental phenomenon on a microscopic scale (parity non-conservation) and macroscopic scale (lack of antimatter in the Universe).

Dr. Wołczek discussed the idea of Svante Arrhenius and--in modern version--that of Crick and Orgel on the extraterrestrial origin of life on the Earth.[37] Biogenic molecules have particular chirality, similar molecules of abiogenic origin have equal amounts of left- and right-handed molecules, as the molecules found in the meteorites.

It is almost certain that only in conditions prevailing on the Earth could life persist and develop until it achieved the highest evolutionary stages.[38] According to present knowledge, the existence of foreign planetary systems seems very probable. Recent computer simulation indicated that only on some planets of stars comprised of a narrow range of stellar spectral classes could life evolve. The rise of intelligent beings could happen only in some biospheres. The possible scarcity of such beings and of their scientific and technical civilizations is understandable in the light of new information given in his paper.[38]

The time of duration of separate stages of planetophysical evolution of the Earth was determined by its mass, its position in the Solar System (the Sun's gravitational field) and the existence of a near-by situated relatively large Moon (its gravitation field and tide effects). Therefore it is possible that the change of one or several of these parameters might cause an adequate shortening or extension of the Earth's individual evolutionary stages deeply influencing the rise and evolution of the biosphere. Conclusion: In the Solar System only the Earth--most favorably situated in a very narrow ecosphere of this system--is a planet on which intelligent life could rise and develop.[37]

The paper Dr. Wołczek presented in Tokyo (1980) during the IAF Congress dealt with the influence of extraterrestrial environments on the possibility of CETI (Communication with Extraterrestrial Intelligence).[39] The author was looking for the influence of the cosmogonical and astrophysical factors on the course of the biological evolution of the Earth-like planets. On these planets should develop the four principal dynamic systems: internal or global (circulation of matter inside of the planet, tectonic and volcanic activity), atmospheric, magnetospheric, and biospheric. These systems could evolve on planets around stars of late F and G types, and some stars of K and M types. As a rule, terrestrial planets seem to appear in the region near the central star at a distance from a few tenths to a few AU. Their masses inferred from the example of the Solar System and from modeling vary from about 0.1 to 5 masses of the Earth, gravity force near 1 g, where g means the gravitational acceleration on the planet's surface, magnetic field screening the atmosphere against charged particles, and the moon determining the proper size and shape of the magnetosphere. All the above factors determine the possibility of the origin and proper evolution of higher organisms. Taking into account these factors one can elaborate the mathematical models of the development of life on the Earth to determine the probabilities of the particular parts of evolution of life on Earth.

The problem of barriers in CETI-SETI (where SETI means Search for Extraterrestrial Intelligence) was discussed in the paper.[40] Human interstellar flight should be realized at very large distances in space and time with relativistic velocities. The distances and velocities are also the barriers, as well as the interaction with the interstellar environment: plasma and dust. Nobody knows the influence of relativistic flight on the human body. There are also the barriers connected with very high costs of CETI, lack of interest, fears of possible dangers caused by mutual contact, or even barriers in mutual understanding.

In this last paper Dr. Wołczek analyzed cosmic influences on the biological evolution and on SETI.[41] He defined the cosmic factors influencing the development of the biosphere and promoting evolution: variation of the radiation flux, both electromagnetic and corpuscular, of the Sun or a particular star, the fall of asteroidal and cometary bodies on the planetary surface, explosion of nearby supernovae, gamma flashes from the galactic nucleus, penetration of the planetary system by clouds of interstellar matter (dust), capture of a moon by the planet and very intensive tide effects.

Sometimes it happens that not one but two or even more of the above factors interact at the same time, changing deeply the course of evolution. Probably strong cosmic factors repeatedly influenced living organisms and their evolutionary path, choosing optimal solutions. Planetary factors act slowly, cosmic factors-rapidly--changing the planetary environment irreversibly. Mankind on the Earth is relatively young. Man has existed probably about four million years. *Homo sapiens* appeared just 110-100 thousand years ago. His civilization exists merely several thousand years. Therefore mankind never experienced the severe stroke of a cosmic factor. Dr. Wołczek thought cosmic factors may stimulate the quest for alien civilization and facilitate our SETI. A highly developed civilization exposed relatively frequently to the dangers of strong cosmic influences may develop protective measures enabling not only survival but also a quick recovery followed by further development.

## LIST OF SCIENTIFIC PAPERS OF DR. OLGIERD WOŁCZEK

### Nuclear Energy in Rocketry

1. Einige Bemerkungen über den sogenannten Annihilationsreaktor und dessen Verwendungsmöglichkeit in den interstellaren Flugen, Bericht über d.VIII Internat. Astronaut. Kongress, Barcelona, 1957, ed. J. Springer, Wien, 1958.

2. Kalte Fusionsreaktoren und deren Verwendungsmöglichkeit zum Antrieb von Raumfahrzeugen, Raketentechnik u. Raumfahrt, 2, Nr.2, 1958.

3. Künstliche Kernenergie Quellen im kosmischen Raum, Proc. IX-th Int. Astronaut. Congress, Amsterdam 1958, ed. Springer-Verlag, Wien 1959, pp.67-87.

### Various Problems, Impact of Astronautics

4. Some remarks on free radicals and their possible use in rocket propulsion, *J.B.I.S.*, 17/1959-60/ pp.133-136.

5. On technical realisation of subgravity and weightlessness, Proc. X-th Int. Astronautical Congress, London, 1959, pp.202-210, ed. Springer-Verlag, Wien, 1960.

6. Die technische Realisation der verringerten Schwerkraft und der Gewichtslosigkeit /Zusammenfassung/. 8 Raketen und Raumfahrttagung 1959 in Cuxhaven.

7. Check altimeter for use in cis-lunar space /in Polish, English and Russian/, *Astronautyka*, 7/1964/ pp.30-32.

8. Znaczenie badań kosmicznych dla nauki i techniki /in Polish: Importance of outer space research for science and technology/, *Zagadnienia Naukoznawstwa* 1 /25/, 1971 pp.32-46.

9. The importance of space research for the future development of science, technology and the civilization of mankind, *Acta historiae rerum naturalium necnon technicarum*, Special issue 8, Prague 1976, It was presented at the Symposium of Int. Cooperation in History and Technology Committee, Kaluga, 7-11, 6, 1976.

10. Astronautics - prerequisite of development of the human civilization and culture, preprint No. IAF-82-421 of IAF Congress in Paris, 26.IX-2.X.1982.

## Matter Evolution; Planetology

11. Rozwój materii w Układzie Słonecznym a astronautyka Post. *Astronautyki*, 6, Nr 2/15/, str. 27-68, 1972.

12. Die Entwicklung der Materie im Weltraum und die Raumfahrt /presented to the XXI Space Congress of the H. Oberth-Gesellschaft in Garmisch-Partenkirchen, 28.9.-1.10.1972/, unpublished.

13. Az urkutatas es az anyag fejlodese a napiendszerben /in Hungarian: Astronautics and the evolution of matter in the Solar System/, presented in Budapest 16/17.10.1972/, published in *Asztronautikai Tajekoztato*, No, 2/27, 23, 1973, pp.23-37.

14. "Pewne ważne zadania Międzynarodowego Laboratorium Marsyjskiego", materiał do dyskusji w Baku 9.10.1973 /materials for discussion during the XXIV IAF-Congress in Baku - 9.10.1973, unpublished/ - "On some important roles of the International Martian Laboratory".

15. Some problems of evolution of the Solar System /I/ /in Polish *Post. Astronautyki*, 7, Nr 2/3 /17/18, 1974 str. 183-193.

16. The evolution of the solar system, and space energy sources and its conversion /in Polish/ /II/, *Post. Astronautyki*, 7, Nr 4 /19/, 1974, p.5-19 /presented to the XXIII Space Congress of the H.Oberth-Gesellschaft in Salzburg, 24-29, VI. 1974/.

17. Unmanned exploration of the Solar System; a critical review and the recommendations for the future, Proc. XXV IAF Congress, Amsterdam, 30.09-5.10.1974, paper 74-048.

18. Badanie Układu Słonecznego przy użyciu urządzeń bessałogowych /in Polish/, *Postępy Astronautyki*, 8, Nr 1 /20/, 1975, pp.77-101.

19. Some future trends in the development of terrestrial planets probes /in Polish/, *Postępy Astronautyki*, 9, Nr 1/24, 1976 pp.7-18 /paper was presented to the IAF Congress in Lisbon, 21-27 Sept., 1975/.

20. Zagadnienie występowania wody na Marsie /in Polish: on the problem of the existence of water on Mars/, *Postępy Astronautyki*, 9, Nr 3/26, 1976, pp.49-71.

21. Contemporary view of terrestrial planets, *Postępy Astronautyki*, 9, Nr 4/27, 1976, pp.26-61.

22. Exploration of planets by remote active means, *Postępy Astronautyki*, 10, Nr 2/29, 1977, pp.59-70 /paper was presented at the XXVII-th IAF-Congress in Anaheim, California, 10-16.X.1976.

23. Terradynamic devices and the exploration of Venus, *Postępy Astronautyki*, 11, Nr 1/32, 1978, pp.21-42.

24. Some problems of investigation of giant planets and their natural satellites, *Postępy Astronautyki*, 12 Nr. 2/37, 1979, 79-88.

25. Missions to asteroids, *Postępy Astronautyki*, 14, Nr 1/2, 1981, p.17-29 and Preprint of IAF-Congress in Munich 17-22 IX.1979.

26. Planety ziemskie jako układy dynamiczne /in Polish: Terrestrial planets as dynamic systems/, *Postępy Astronautyki*, 13 Nr 2 /1980/, pp.19-40.

27. Układy dynamiczne jako czynniki warunkujące strukturę przemiany i ewolucję planet - na przykładzie planet ziemskich /2/ in Polish: Dynamic systems as factors determining the structure, transformations and evolution of planets exemplified by the terrestrial planets /2/, *Postępy Astronautyki*, 13, /1979/ pp.81-108.

28. Układy dynamiczne jako czynniki..../3/in Polish: Dynamics systems as factors..../3/ *Postępy Astronautyki*, 13, Nr.4 /1980/, pp.21-48.

29. Dynamic systems as factors determining... *Adv. Space Res.*, vol. 1 /1981/ pp.217-221

30. New conceptions of unmanned planetary exploration /I/ *Postępy Astronautyki*, 14, No.1/2 /1981/, pp.31-37, Preprint IAF-80, G-293 of IAF Congress in Tokyo, 1980.

31. New conception of unmanned planetary exploration /II/, Extra-solar planetary systems, preprint IAF-81-206 of IAF Congress in Rome, 6-12. Sept. 1981.

32. Some novel aspects of cometary research, preprint No. IAF-8-82-196 of Congress in Paris, 26.IX-2.X.1982.

## Life in the Universe

33. Wyniki doświadczań biologicznych w próbnikach "Viking" a możliwośc występowania życia na Marsie. /in Polish: Results of biological experiments in the "Viking" landers and the possibility of existence of life on Mars/, *Postępy Astronautyki*, 10, Nr. 4/31, 1977, pp.51-90.

34. Occurrence of life and rise of intelligent beings as stages of evolution of cosmic matter, *Postępy Astronautyki*, 10, Nr 1/28, 1977, pp.67-86.

35. Möglichkeiten der Lebensentwicklung ausserhalb der Erde, Bibliothäk des Geschwadearztes /DDR/, Sonderheft Nr. 1/78, Teil 1, Nationale Volksarmee, Kommando der Luftstreikrafte u. Luftverteidigung, 1978, p. 48-54.

36. Chirality of compounds essential for the existence of life, and the possibility of its evolution in the Universe, *Postępy Astronautyki*, 11 nr 4/35, 1978, pp.53-80.

37. Problemy pozaziemskiego pochodzenia życia /in Polish: Some problems of the extraterrestrial origin of life/, *Postępy Astronautyki*, 12, Nr 1/36, 1979, pp.91-101.

38. On the possibility of appearance and development of extraterrestrial intelligent beings, Preprint of the XXX IAF Congress in Munich, 17-22.IX.1979, pp.1-7.

39. The influence of evolution of extraterrestrial environments on the possibility of CETI, preprint No. IAA-53 of XXXI IAF Congress in Tokyo, 21-28.IX.1980.

40. Problems of barriers in the development of SETI and CETI, Preprint No. IAA-81-303 of XXXII IAF Congress in Rome 6.12.IX.1981.

41. Impact of biological evolution on SETI - cosmic influences, preprint No. IAA-82-270 of XXXIII IAF Congress in Paris, 26.IX-2.X.1982.

# APPENDIX

# APPENDIX

# BIOGRAPHICAL SKETCHES OF THE AUTHORS*

**Chae, Yeon-Seok:** Born in 1951, in Choongbook, Korea, he received a B.S. in physics (1975) and a M.E. in mechanical engineering (1977) at the Kyung Hee University, Seoul, Korea. Chae was a graduate student at the University of Mississippi, U.S.A., when he wrote the treatise presented here. He received a M.S. (1984) and a Ph.D. (1987) from the University of Mississippi's Department of Aerospace Engineering. After a summer faculty position at the NASA Lewis Research Laboratory in Cleveland, Ohio, Chae returned to Korea where he is now a senior researcher in space engineering at the Institute of Space Science and Astronomy, Taejon, Korea. He is the author of two books: *Rockets and Space Flight* (Korea, 1972) and *Early Firearms in Korea* (Korea, 1980) as well as six treatises on rockets and fluid flow.

**James A. Dewar:** Majored in the history of science and technology and received his doctorate from Kansas State University, U.S.A., in 1974. He is presently working at the U.S. Department of Energy in formulating and implementing nuclear weapon arms control and nonproliferation policies.

**Burton I. Edelson:** Received his B.S. degree from the U.S. Naval Academy (1947) and Ph.D. from Yale University (1960). He served for twenty years as a naval officer at sea and in research and engineering positions, including the Office of Naval Research and the National Aeronautics and Space Council. From 1968 to 1982, Edelson held executive positions at the Communications Satellite Corp., including Director of COMSAT Laboratories and the Corporation's central R&D facility; Vice-President for Systems and Engineering, and Senior Vice President. From 1982 to 1987, Edelson was Associate Administrator for Space Science and Applications of the National Aeronautics and Space Administration (NASA). Since leaving NASA, he has been a Fellow of the Foreign Policy Institute of the School of Advanced International Studies of the Johns Hopkins University where he specializes in international science and technology policy.

**Kristan R. Lattu:** Graduated from the University of California, Riverside, U.S.A., and is a staff engineer with the Jet Propulsion Laboratory of the California Institute of Technology. She worked in flight operations with NASA's Voyager project during the Jupiter and Saturn encounters, developed operations plans for the Galileo project, supported early design studies for instrument integration, test, and servicing of the Earth Observing Systems polar orbiting platforms, and is (1989)

---

\* **Editor's Note:** An attempt was made to obtain the biographies of all authors. Those who responded appear here in this appendix.

supporting joint studies for payload integration and operations on the Space Station *Freedom*.

**István György Nagy:** Born in 1905 in Budapest, Hungary. He holds a master of mechanical engineering degree and was active in the field of military technology. Nagy was one of the founders of the Hungarian Astronautical Society and has presented papers at several IAF Congresses. He is the author or co-author of fifteen books dealing with military technology (especially rocketry) and astronautics.

**Frederick I. Ordway, III:** Born in New York City, U.S.A. in 1927, he received a B.S. in geosciences from Harvard University in 1949 and did postgraduate work at the University of Paris, Air University, Industrial College of the Armed Forces, and the Universities of Algiers, Innsbruck, and Barcelona. After several years of geological work in Venezuela, Ordway came to the United States and has been working in the fields of rockets and energy in industry and government. He has written over 300 articles and 15 books on rockets and astronautics, and has also served as an editor and consultant for many publications and activities on rockets and space. He is (1989) working with the U.S. Department of Energy, Washington, D.C.

**L. R. Shepherd:** Leslie Robert Shepherd was born in Glamorgan, United Kingdom, in 1918. He received a degree in physics from University college, London in 1940. He was a graduate student at St. Catherine's College, Cambridge, where he worked in the Nuclear Physics Department of the Cavendish Laboratory and completed his Ph.D. in 1949. From 1948 to 1955, Shepherd worked on the development of fast reactors at the Atomic Energy Research Establishment. He was an original member of the team at Harwell which pioneered the concept of the High Temperature Gas-Cooled Reactor (HYGR) and between 1959 and 1976, worked on the Dragon Project in which 13 European countries took part. This project designed, constructed and operated the world's first HTGR. Shepherd headed the R&D part of the project for nine years and from 1968 to 1976 was its Chief Engineer. He joined the British Interplanetary Society (BIS) in 1935 and has served on its Council since 1946. He served three times as Chairman/President (1953-56, 57-60, 66-67). He served as President of the International Astronautical Federation (IAF) in 1956/57 and 1962.

**M. Subotowicz:** Mieczyslaw Subotowicz was born in Wilno, Poland, in 1924. He received a M.Sc. and Ph.D. at the Institute of Physics, M. Curie-Sklodowska University (UMCS) in Lublin, Poland. He served as Dean of the Mathematics, Physics & Chemistry faculty of UMCS from 1966 to 1975 and head of the Experimental Physics Department from 1975 to present (1989). He is a member of many Polish and international scientific organizations. He has authored 16 books, more than 400 scientific papers and more than 300 popular-scientific articles. His current interests include solid state physics, nuclear physics, space physics (interstellar communication), history of science, and the social consequences of the development of science.

**G. V. E. Thompson:** Gordon Vallins Elliot Thompson is a graduate of the Imperial College of London, the Open University, and Brunel University. After working as an industrial chemist cum chemical engineer, he served as senior information officer to the British Non-Ferrous Metals Research Association from 1948 to 1961, when he became a translator for (and translations editor of) the Russian Journal of Inorganic Chemistry (published by the Royal Society of Chemistry). He has been a member of the Council of the British Interplanetary Society (BIS) since 1946 and was editor of the Society's Journal from 1957 to 1964, a Vice-President for six years, and President of the BIS 1979-82. He is currently (1989) Secretary of the Macular Disease Society.

**Robert C. Truax:** Robert Truax is an American pioneer in the development of rockets, missiles, and space vehicles. He began his career in rockets in 1937 as a midshipman at the U.S. Naval Academy and, until he retired from the Navy as a Captain in 1959, he was connected in one way or another with about every Navy rocket engine project during that period. Some of this work is described in his treatise presented here. After World War II, Truax headed rocket propulsion in the Navy Bureau of Aeronautics where he supervised development of the propulsion systems for experimental supersonic aircraft. In the 1950s, Truax headed work on surface launched missiles in the Navy involving the Regulus missile and studies of fleet-launched missiles leading to the Polaris and Trident. During this time he added a degree in aeronautical engineering and a masters degree in mechanical engineering to his Navy academy engineering degree. Also during the 1950s, Truax was loaned to the U.S. Air Force and headed work leading to the Thor missile. After military retirement, Truax continued rocket research and development in the private sector, heading advanced development at Aerojet-General Corporation for six years. In 1973, he formed his own company for the commercialization of space. He has been awarded many honors for his contributions to rocketry and astronautics.

**Frank H. Winter:** Born in London, England, in 1942, Winter emigrated to the United States in 1951. After attending various public schools and colleges in New York City and Los Angeles, California, he served in the U.S. Air Force from 1964 to 1968. Stationed in Spain from 1966 to 1968, Winter served as a military journalist and feature editor. He won the Robert Goddard essay contest in 1966 and was a co-winner in 1970. He began working for the Smithsonian Institution in 1969 and in 1977 received a B.A in history from the University of Maryland. Winter is currently (1989) Curator, Rocketry, at the National Air and Space Museum. He is the author of numerous articles and several books on rockets and space.

# INDEX

# INDEX TO ALL AMERICAN ASTRONAUTICAL SOCIETY PAPERS AND ARTICLES 1954 - 1985/86

This index is a numerical/chronological index (which also serves as a citation index) and an author index. (A subject index volume will be forthcoming.)

It covers all articles that appear in the following:

*Advances in the Astronautical Sciences* (1957 - August 1986)

*Science and Technology Series* (1964 - September 1986)

*AAS History Series* (1977 - 1986)

*AAS Microfiche Series* (1968 - August 1986)

*Journal of the Astronautical Sciences* (1954 - 1986)

*Astronautical Sciences Review* (1959 - 1962)

If you are in aerospace you will want this excellent reference tool which covers the first 30 years of the Space Age.

Numerical/Chronological/Author Index in two volumes, Library Binding (both volumes) $95.00; Soft Cover (both volumes) $80.00; Volume I (1954 - 1978) Library Binding $40.00; Soft Cover $30.00; Volume II (1979 - 1985/86) Library Binding $60.00; Soft Cover $45.00.

**Order from Univelt, Inc., P.O. Box 28130, San Diego, California 92198.**

# NUMERICAL INDEX*

| | |
|---|---|
| **Volume 12** | **AAS HISTRORY SERIES,** *HISTORY OF ROCKETRY AND ASTRONAUTICS*, (IAA History Symposia, Volume 7) |
| | (Proceedings of the Seventeenth History Symposium of the International Academy of Astronautics, Budapest, Hungary, 1983) |
| AAS 91-281 | A Study of Early Korean Rockets (1377-1600), Chae, Y. S. |
| AAS 91-282 | Leonhard Euler's Importance for Aerospace Sciences - On the Occasion of the Bicentenary of his Death, W. Schulz |
| AAS 91-283 | The Founding of the Jet Propulsion Research Institute and the Main Fields of its Activity, B. V. Rauschenbach |
| AAS 91-284 | The British Interplanetary Society: The First Fifty Years (1933-1983), G. V. E. Thompson, L. R. Shepherd |
| AAS 91-285 | Liquid Propellant Rocket Development by the U.S. Navy during World War II: A Memoir, R. C. Traux |
| AAS 91-286 | Some Vignettes from an Early Rocketeer's Diary: A Memoir, B. Smith as told to F. I. Ordway, III |
| AAS 91-287 | Contribution of the Romanian Inventor Alexandru Churcu to the Development of Theoretical and Practical Reactive Motion in the 19th Century, F. Zăgănescu, R. Burlacu, I. M. Stefan |
| AAS 91-288 | Communication Satellites: The Experimental Years, B. I. Edelson |
| AAS 91-289 | Project Rover: The United States Nuclear Rocket Program, J. A. Dewar |
| AAS 91-290 | A Comparative Study of the Evolution of Manned and Unmanned Spaceflight Operations, K. Lattu |
| AAS 91-291 | Reaction Motors Division of Thiokol Chemical Corporation: An Operational History, 1958-1972 (Part II), F. I. Ordway, III |
| AAS 91-292 | Reaction Motors Division of Thiokol Chemical Corporation: A Project History, 1958-1972 (Part III), F. H. Winter |
| AAS 91-293 | Pages from the History of the Hungarian Astronautical Society, I. G. Nagy |
| AAS 91-294 | United States Space Camp at the Alabama Space and Rocket Center, E. O. Buckbee, L. Sentell |
| AAS 91-295 | A Life Devoted to Astronautics: Dr. Olgierd Wolczek (1922-1982), M. Subotowicz |

---

\*  IAA history papers presented in 1983. AAS numbers have been assigned for identification purposes.

# AUTHOR INDEX*

Buckbee, E. O., AAS 91-294, His v12, pp209-214

Burlacu, R., AAS 91-287, His v12, pp85-91

Chae, Y. S., AAS 91-281, His v12, pp3-16

Dewar, J. A., AAS 91-289, His v12, pp109-124

Edelson, B. I., AAS 91-288, His v12, pp95-108

Lattu, K., AAS 91-290, His v12, pp125-136

Nagy, I. G., AAS 91-293, His v12, pp203-207

Ordway, III, F. I., AAS 91-286, His v12, pp69-84; AAS 91-291, His v12, pp137-173

Rauschenbach, B. V., AAS 91-283, His v12, pp31-36

Schulz, W., AAS 91-282, His v12, pp19-28

Sentell, L., AAS 91-294, His v12, pp209-214

Shepherd, L. R., AAS 91-284, His v12, pp37-55

Smith, B., AAS 91-286, His v12, pp69-84

Stefan, I. M., AAS 91-287, His v12, pp85-91

Subotowicz, M., AAS 91-295, His v12, pp217-231

Thompson, G. V. E., AAS 91-284, His v12, pp37-55

Traux, R. C., AAS 91-285, His v12, pp57-67

Winter, F. H., AAS 91-292, His v12, pp175-201

Zăgănescu, F., AAS 91-287, his v12, pp85-91

---

\* For each author the paper number is given. AAS numbers for these IAA history papers presented in 1983 have been assigned for identification purposes. The page numbers refer to Volume 12, AAS History Series.